The Innovation Delusion

The Innovation Delusion

HOW OUR OBSESSION WITH THE NEW HAS DISRUPTED THE WORK THAT MATTERS MOST

Lee Vinsel and Andrew L. Russell

CURRENCY · NEW YORK

Published in the United States by Currency, an imprint of Random House, a division of Penguin Random House LLC, New York.

CURRENCY and its colophon are trademarks of Penguin Random House LLC.

Parts of chapters 7 and 11 appeared as Lee Vinsel, "Fighting for the Right to Repair Our Stuff," *The American Conservative,* July 23, 2019.

Library of Congress Cataloging-in-Publication Data
Names: Vinsel, Lee, author. | Russell, Andrew L., author.
Title: The innovation delusion / Lee Vinsel and Andrew L. Russell.
Description: First edition. | New York: Currency, [2020] | Includes bibliographical references and index.
Identifiers: LCCN 2020012053 (print) | LCCN 2020012054 (ebook) | ISBN 9780525575689 (hardcover; alk. paper) | ISBN 9780525575696 (ebook)
Subjects: LCSH: Technological innovations—Economic aspects. | Technological innovations—Social aspects. | New products. | Social responsibility of business. | Sustainable development.
Classification: LCC HC79.T4 V56 2020 (print) | LCC HC79.T4 (ebook) | DDC 658.4/063—dc23
LC record available at https://lccn.loc.gov/2020012053
LC ebook record available at https://lccn.loc.gov/2020012054

Printed in Canada on acid-free paper

currencybooks.com

9 8 7 6 5 4 3 2 1

First Edition

Book design by Debbie Glasserman

Never set out to innovate, because more horror is
done with that goal in mind than any other.

—CHARLES EAMES, LEGENDARY DESIGNER

Another flaw in the human character is that everybody
wants to build and nobody wants to do maintenance.

—KURT VONNEGUT

Contents

The Innovation Delusion

Part One

CHAPTER ONE

The Problem with Innovation

For want of a nail the shoe was lost.
For want of a shoe the horse was lost.
For want of a horse the rider was lost.
For want of a rider the message was lost.
For want of a message the battle was lost.
For want of a battle the kingdom was lost.
And all for the want of a horseshoe nail.

—"FOR WANT OF A NAIL," UNDATED POEM

The explosions started at 8:00 A.M. with a spark from a bookstore furnace. Gas had leaked from a corroded local storage tank and into the city sewers overnight, and the cloud of vapor wound its way around the system before escaping through floor drains of downtown stores. The explosions rocked four buildings in all. Nobody was hurt, but authorities evacuated twenty thousand people from a thirteen-block area. It was an unpleasant start to a cold April morning in Saint John, a small town in the Canadian province of New Brunswick.[1]

On that day in 1986, four of the buildings directly above the leak were badly damaged. But one of their neighbors, located within the same radius, was spared. Why?

The person who knows the answer—Heidi Overhill—let us in on the secret. Her late father, T. Douglas Overhill, ran an engineering consulting firm specializing in preventive maintenance. His favorite poem, which we quote above, was "For Want of a Nail," a paean to the far-reaching consequences of neglected maintenance. One of Overhill's clients, the owner of an office building in Saint John, had been following a plan Heidi's father designed for maintaining the property. Heidi described it to us in detail: "One of the scheduled tasks was to pour a bucket of water down each of the basement floor drains. Floor drains tend to dry out, and when there is no water in the S-shaped traps in the drainpipes, bad smells [and explosive gases] from the sewers can leak up through them." The fix for this problem is simple—"a bucket of water every now and then will seal the trap and make the basement smell better."

On the day of the explosions, the surviving building belonged to Overhill's client, who had recently poured a bucket of water into the floor drain to seal it. But the owners of the neighboring stores that were damaged had followed no such plan.

In real life, as in Overhill's favorite poem, the kingdom was lost—and all for the want of maintenance.

Do you ever get the feeling that everyone around you worships the wrong gods? That, through fluke or oversight, our society's charlatans have been cast as its heroes, and the real heroes have been forgotten?

In a 2009 interview, Facebook CEO Mark Zuckerberg, reflecting on his young company's success, shared what has become a mantra for our times: "One of the core values of Facebook is 'Move fast and break things.' Unless you are breaking some stuff you are not moving fast enough."[2] Rapid growth is the sine qua non of the digital economy—just ask anybody who has owned stock in Google, Apple, Facebook, or Amazon. New features draw new users and more revenue from advertisers and subscribers, which helps companies secure more funding and hire more people.

Digital upstarts like Facebook succeed when they displace incumbents; that is why Zuckerberg was comfortable with the costs of taking risks. "One of the trade-offs that we made," he later remarked, "was we tolerated some defects in the product." This tactic works in the digital economy, where users are accustomed to beta releases and flaky connections, and the costs of fixing broken code pale in comparison to the costs of fixing a physical product, such as a car with faulty airbags or a bookstore with dried-out floor drains. In other words, "move fast and break things" is something more than a juvenile crack from a CEO who was twenty-five years old at the time he said it. It's a business strategy, an ethos that applies equally to product development and to Facebook's aggressiveness in buying out potential rivals, such as WhatsApp and Instagram.

Zuckerberg wasn't alone in this outlook. At least since the dot-com crash of 2001, CEOs, entrepreneurs, and business school professors flouted common sense with buzzwords like "disruptive innovation" and "creative destruction," not to mention the imperative to "fail faster [to] succeed sooner."[3] This approach quickly became recognized as the "start-up" mentality, and *innovation* was its prime directive—a demand for rapid growth that disrupts the comfortable incumbents of the status quo. To be sure, this innovation mindset led to some amazing things. Sixteen years after being launched, Facebook has more than two billion users around the world. Billions more would have a hard time functioning without constant access to Google or an iPhone.

As business leaders embraced this worldview, its effects spilled out beyond the economy. We adjusted our values, even our vision of democracy, to be suitably deferential to the gods of Silicon Valley. We tolerated increasing amounts of "screen time" for our children and pledged our attention to addictive apps. A 2018 Georgetown University survey found that Americans trust Amazon and Google more than local, state, or federal government.[4] In early 2016, an op-ed in *The Wall Street Journal* even floated the idea of a new political party that could bring "radical disruption" to "Establishment America."[5] The leaders of this movement could come from Silicon Valley—perhaps Mark Zuckerberg, Sheryl Sandberg, or another of its heroes—and they could call it the "Innovation Party." After all, the essay concluded, "Who is against innovation?" The new political party failed fast, never moving past the op-ed phase.

Novelty is at the core of American identity. (How many of our cities have names that begin with "New"?) Since the sixteenth century, we've been pushing stubbornly past "frontiers" of all kinds to reap the bounties of natural resources, political autonomy, and scientific progress. In the twenty-first century, our new digital gadgets were self-evident emblems of the superiority of the innovation mindset. The companies that made these gadgets and their "killer apps" grew until they were the most highly valued corporations in world history. Their lush corporate campuses became coveted destinations for college graduates. Their executives became icons. A nation mourned in 2011 when Apple's CEO Steve Jobs died. Serial entrepreneur Elon Musk was named among 2019's most admired people in America—ahead of Pope Francis and the Dalai Lama, but a significant distance behind Barack Obama and Donald Trump, America's disrupter in chief.[6]

And so, Americans went all in on innovation. Businesses created new positions like chief innovation officer and "Innovation Evangelist." Universities invested millions of dollars to build flashy new Innovation Centers, and philanthropists supported ambitious proposals for transforming some of our most basic cultural institutions. Schools at the K–12 level "disrupted" education by introducing laptops and tablets into the classroom and seeking to instill characteristics like "grit," entrepreneurialism, and "Design Thinking" in their students. Millennials in the job market reported feeling worthless and burned out if their creative exploits fell short of their own expectations or those of people they followed on Instagram. The result of all this change is dubious—in most cases, advocates cannot show that the efforts to stoke innovation have delivered on their promises. But that hasn't stopped Americans from upending centuries of tradition in the name of newfangled fads.

The entrepreneurs and investors of Silicon Valley have profited from software, and this success has given them the capital and confidence to branch out into other fields. But while Zuckerberg's advice to "move fast and break things" is still considered good counsel for web designers and app builders[7]—professions where profit margins are high, the costs of failure are low, and venture capital is plentiful—it turns out that "move fast and break things," and the innovation mindset more generally, can be lousy guidance for anyone who builds or designs *actual* things.

In 2016, reviewers celebrated the Samsung Galaxy Note 7 as a "beautiful" validation of Samsung's "innovation strategy"; that is, until hundreds of customers began to complain about burns and property damage caused by the phone's exploding battery. A Miami bridge praised for its "innovative" design killed six people when it collapsed onto a six-lane highway in 2018.[8] And Elizabeth Holmes, who in 2003 founded the blood-testing start-up Theranos at age nineteen, moved fast—raising more than $700 million from investors and achieving a $10 billion valuation for her company. But Theranos also broke things, namely, laws protecting investors from the fraudulent, dangerous claims about the company's "revolutionary" technology. When digital-age companies encounter old problems in their new ventures in the material world—logistics, manufacturing, consumer tastes, societal norms and regulations, and traditional dynamics of supply and demand—they consistently flounder.

How do people with terrible ideas find hundreds of millions of dollars to help them fail fast? Well, hubris surely has something to do with it. It's easy to poke fun at ventures like Juicero—the $700 juicing machine that raised $120 million in venture capital before users discovered they could squeeze the machine's pre-juiced fruit packets by hand. It would be tempting, too, to focus only on the handful of (mostly white male) serial entrepreneurs who trip over their shoelaces through a world of wealth, ambition, and bad advice. But these problems have metastasized far beyond the confines of Silicon Valley, permeating America's economy—and, arguably, global culture.

We are writing this book because we are sick of hearing about what's good for Silicon Valley, and what the innovating classes think is good for us. We think it's time to refocus on what's healthy for the vast majority of workers, for the businesses that aren't at the cutting edge of digital transformation, and for all of us who don't want to be subject to the whims of a few out-of-touch billionaires.

For the past several years, we have researched how the gospel of innovation has affected transportation, computing, and other technological systems, while reflecting on the overlooked fields of infrastructure and maintenance. We have discovered a culture that seeks to apply the wrong lessons from the digital world to the physical world, a culture whose conception of technology reflects an unholy marriage of Silicon Valley's conceit with the worst of Wall Street's sociopathy. These reflections—and an attempt to describe a healthier way forward for technology and society—have become our passions, and they are the subjects of this book.

THE DIFFERENCE BETWEEN INNOVATION AND INNOVATION-SPEAK

For the rest of the book to make sense, there is a distinction that we must make. The distinction has to do with the way we talk about change—specifically, innovation. There is *actual innovation,* the profitable combination of new or existing knowledge, resources, and/or technologies. The Austrian economist Joseph Schumpeter argued that innovation is the motive force of economic change, capitalism, and indeed history itself. But genuine innovation is quite distinct from *innovation-speak,* a breathless dialect of word salad that trumpets the importance of innovation while turning that term into an overused buzzword. As we will see, the world we actually inhabit, including the technologies we use and need, is a very different place from the world described to us by marketing departments and CEOs—replete with the technologies they've convinced us to buy and rely on.

Innovation, at its core, is change that can be measured because it generates profits. A recent iconic example is Apple's iPhone: It generated tremendous profits not because it was a new invention but because it combined a variety of existing functions within a customer-friendly design. Other prominent twentieth-century innovations include hybrid vehicles, virtual reality goggles, cochlear implants, functional MRI scanners, and genetic testing. Older innovations make up the fabric of our daily lives: electric power, reinforced concrete, the internal combustion engine, and synthetic materials like Teflon and neoprene.

Because innovation is such a flexible term—and because its success is followed by profit—its promoters have wrapped the concept in promises about its future impact. "The Segway will change the world!" "We're entering the era of the paperless office!" "The telegraph/airplane/Internet will usher in a new era of world peace!" And so on. We call this hype innovation-speak. Unlike actual innovation, which is tangible, measurable, and much less common, innovation-speak is a sales pitch about a future that doesn't yet exist.

Innovation-speak is fundamentally dishonest. While it is often cast in terms of optimism, talking of opportunity and creativity and a boundless future, it is in fact the rhetoric of fear. It plays on our worry that we will be left behind: Our nation will not be able to compete in the global economy; our businesses will be disrupted; our children will fail to find good jobs because they don't know how to code. Andy Grove, the founder of Intel, made this feeling explicit in the title of his 1996 book *Only the Paranoid Survive*. Innovation-speak is a dialect of perpetual worry.

At a deeper level, innovation-speak is built on the hidden, often false premise that innovation is inherently good. To cite an (admittedly extreme) example, more than one academic article has examined how crack cocaine "disrupted" the market for hard drugs in the 1980s.[9] Similarly, the products and business strategies that undergird our current opioid crisis—including shipping millions of pills to small Appalachian towns and marketing the drugs aggressively to physicians—fit the definition of an innovative business model. They generate profit by carving out new distribution channels and creating new customer demand, as detailed in a 2009 article on the overpromotion and overprescription of OxyContin published in the *American*

Journal of Public Health: "Although OxyContin has not been shown to be superior to other available potent opioid[s] . . . by 2001 it had become the most frequently prescribed brand-name opioid in the United States for treating moderate to severe pain." The author described the promotion and marketing of the drug as a "commercial triumph, public health tragedy."[10]

The ideology of change for its own sake is a recipe for disaster in the wrong hands. *Fortune* magazine named Enron the most innovative company in America from 1996 to 2001, before the energy giant's shady accounting practices came to light. Elsewhere, lawmakers have applied the "start-up mentality" to education with dreadful results, unleashing a flood of for-profit schools (such as the fraudulent Trump University) and the erosion of funding for public education. The warping of truth and democracy at the hands of conspiracy theorists and foreign governments has become one of the most significant political developments of our time. Yet no one delivers sunny TED Talks on the disruptive innovation of Alex Jones and the Russian intelligence agencies.

Our point here is that many people assume that innovation itself is a good, when in fact it can never be more than a means to an end. In chapter 2, we show that innovation often functions as a proxy for values perceived to be lacking in society—the most common example being when someone proposes techno-solutions to profound social problems. (To quote two examples: "Can an App Solve Racism? This Entrepreneur Says It Can" and "5 iPhone Apps That Help Fight Poverty.")[11] Innovation is sometimes used as a stand-in for practical values like efficiency and convenience or altruistic values like kindness and tolerance. In any case, innovation alone won't take us to where we want to go. To create a prosperous society centered on human flourishing, we will need to make sure that all citizens have access to basic goods, including modern infrastructure; that the people who take care *of* our society are adequately compensated and cared *for;* and that we allocate enough resources to preserve the physical structures and wealth that we have already created, things that can become degraded and lose value and efficacy if neglected.

To be clear: Innovation is important. It has played an essential role in economic growth and improved quality of life—a fact that the two of us have personally experienced in recent visits to hospital delivery rooms, cancer treatment centers, and even our local Apple Stores. We can even tolerate smidges of innovation-speak where it's appropriate, such as the push to develop vaccines and treatments for the COVID-19 virus. And while we disapprove of the way in which start-ups and consultants have distorted terms like "disruption" to promote their ambitions, we appreciate their willingness to tackle some of the world's great challenges in education, healthcare, and the alleviation of poverty.

But much of what passes for innovation is actually innovation-speak. In recent years, economists have noted that the rate of innovation has *decreased* since about 1970.[12] To put it another way, there's no evidence that actual innovation or technological change has increased during the period when everyone started talking about innovation. At its most extreme, innovation-speak actively devalues the work of most humans, especially those who do the dirty work that keeps our technological civilization running. And, as we will see, it fails to capture the essence of human life with technology—where maintenance and reliability are far more valuable than innovation and disruption.

MAINTAINING THE THINGS THAT MATTER MOST

Think about the importance of maintenance in everyday life, starting with your daily commute. Could you get to work if roads and bridges weren't kept in good condition? Or, if you prefer walking, how frustrated would you get if the sidewalks were covered in filth or hadn't been cleared of snow? Maintenance matters. This is why, when the New York City subway breaks down, the MTA's Twitter account tries to reassure stranded riders that "maintainers are en route." We suspect straphangers would find little comfort in hearing "innovators are en route."

Old technologies aren't the only ones that require maintenance. Our culture celebrates software and digital technology as realms of cutting-edge development, but most of the work invested in these state-of-the art operations involves simply keeping them going. Consider what happens when software malfunctions: a call drops; messages disappear; photos vanish; projects and data are lost. The task of diagnosing the problem can test patience, strain marriages, and end with smashed screens. Luckily, there are legions of people to help. Some are paid—think of the help-desk staff at your workplace or the patient souls at the Apple Store's Genius Bar—and others volunteer their time to fix bugs or patch holes in open-source software.

Despite the importance of maintenance, we often fall asleep on the job. Businesses, homeowners, governments, and other groups responsible for public infrastructure often respond to the high costs of maintenance by ignoring them. Deferred maintenance can impose significant and even tragic costs. On August 14, 2018, the Morandi Bridge collapsed in Genoa, Italy, killing forty-three people and leaving six hundred homeless. When the bridge opened in 1967, the Italian press boasted about how it wouldn't require upkeep: "The bridge's concrete structure won't need any maintenance. Neither will its stayed cables, which are protected from atmospheric agents by their concrete vest."[13] They sold the illusion of a maintenance-free future. When investigators sifted through the multiple factors that caused the bridge to collapse, they found that some of its parts had not been maintained for twenty-five years.

In the chapters that follow, we'll show how the innovation mindset has led to a devaluation of maintenance and care, with disastrous results. We'll meet lawyers, teachers, and engineers who have been told they need to be more innovative—even though they know that their success in many ways depends upon *resisting* the pressures to "fail fast" or "move fast and break things." We are fascinated by their acts of resistance and how their attempts to maintain their integrity and do their jobs shed light on a different way forward.

In some ways, maintenance is the opposite of innovation. It is the practice of keeping daily life going, caring for the people and things that matter most to us, and ensuring that we preserve and sustain the inheritance of our collective pasts. It's the overlooked, undercompensated work that keeps our roads safe, our companies productive, and our lives happy and secure.

In other ways, however, maintenance and innovation can be nicely aligned. For example, Corgibytes is a company that specializes in "software remodeling," or helping companies clean up, organize, stabilize, streamline, and nurture the software and code they use for product development, security, and daily operations. Andrea Goulet, the company's CEO, launched Corgibytes in 2009 with her cofounder M. Scott Ford; by 2020 they had nearly twenty people on staff. They have an infectious enthusiasm for their work, and they've given their employees perks like the ability to work from home, flexible hours, and generous benefits for health insurance and professional development.

Cultural expectations—including senses of cleanliness, order, and duty—play an enormous role in determining what kinds of upkeep we choose to do. In the pages that follow, you'll discover dozens of companies like Corgibytes that do not view maintenance as either drudgery or a reactionary force but instead as a good option that taps into values whose time has come—empowered workers, family-friendly policies, environmental sustainability, and economic security—and contributes to the profitability of their bottom line. We hope that our readers will discover, as we did, that if we need role models for our children and leaders for our troubled age, we would do well to look past the icons of Silicon Valley and focus instead on the maintainers.

Because technology permeates nearly every aspect of our lives, we believe it's essential to find a more nuanced and holistic way of thinking about it—its creation, use, and demise. Even Mark Zuckerberg demonstrated some capacity for revising his own views when he announced in 2014 that Facebook would replace "move fast and break things" with a new mantra: "Move fast with stable infrastructure."[14] Although we would hesitate to hold up Facebook as a positive model for much of anything right now, as regulators scrutinize the network's cavalier approach to user privacy and surveillance, we see the change in its slogan as a reason for hope. If the originator of "move fast and break things" is willing to adopt a worldview that acknowledges the complexity of technological change, perhaps our lawmakers and business leaders can follow his example.

A BRIGHTER FUTURE

We started down the path that led to this book in 2014, when the biographer Walter Isaacson published *The Innovators: How a Group of Hackers, Geniuses, and Geeks Created the Digital Revolution*. Foremost among the issues we had with the book was the emphasis Isaacson placed on the shiny and new while neglecting the ordinary, gritty nature of life with computers. Andy proposed that we counter *The Innovators* with a volume of our own, with a title like *The Maintainers: How a Group of Bureaucrats, Standards Engineers, and Introverts Created Technologies That Kind of Work Most of the Time*. We began playing with this idea online, in blog posts and via tweets, and to our surprise it took on a life of its own. Members of our scholarly community, historians and social scientists who study technology, encouraged us to do more with it.[15]

In April 2016, we held a conference titled "The Maintainers" and published an essay, "Hail the Maintainers," in *Aeon.* Then something even more surprising happened: An admittedly nerdy conversation among scholars of technology went viral. Both news of the conference and the essay were picked up by mainstream outlets like *The Atlantic, The Guardian, Le Monde,* and the Australian Broadcasting Corporation. We started receiving emails from people in Africa and Russia, and invitations to speak in places like Brussels and New Zealand. *Freakonomics Radio* produced an episode around the idea, and we were invited to do an interview on NPR and write an op-ed for *The New York Times*.

Since that time—with the help of many, many others—we have held two more conferences and built The Maintainers into a global, interdisciplinary community that examines maintenance, repair, infrastructure, and the ordinary work that typically is forgotten. Our email list—which you can join—includes people from all kinds of backgrounds: industrial consulting, libraries and archives, university and nonprofit administration, open-source software and legacy code maintainers, philanthropies, artists and designers, start-up founders and employees, federal agencies, right-to-repair advocates, labor activists, and many more.

Over the past few years, we heard the same story again and again from the people who reached out to us: that they are sick to death of vapid chatter about innovation; that they no longer believe (if they ever did) that technology alone can solve our most important problems; and that they believe our obsession with the new undermines and devalues the mundane labor people do all around us, on which our lives depend each and every day. The book you are now holding is the result of our thinking, our research, and, most important, information we have gleaned from others over the past six years. In the following pages, you will learn about the members of The Maintainers community and the work they are doing to better our world.

We are deeply concerned about what our society is doing in the name of innovation, disruption, and the reckless mentalities of "fail fast" and "move fast and break things." *The Innovation Delusion* is our attempt to expose the pathologies of innovation-speak, and to help you think in a different way about our technological civilization. We hope that you will find this honesty and clarity refreshing.

Parts 1 and 2 of *The Innovation Delusion* document the damage that has already occurred. We start in the past: How did our society come to hold innovation in such high regard, and why did digital technologies pull our attention away from technology in the broader sense? We then assess the damage caused by the Innovation Delusion in three different spheres: society as a whole; specific organizations such as businesses and schools; and the lives and careers of individuals. We will show that the key issue is our relationship with *technology*—where muddy thinking and poor leadership have combined to create disastrous results.

Part 3 will sketch a healthier and more productive approach.

What would the world look like if we focused more on fixing things instead of throwing them away; if we learned to rely on and respect the everyday technologies that we take for granted rather than worship the new? We'll see this maintenance-oriented mindset at work in Japan and the Netherlands; in software companies and the U.S. Air Force; in libraries, hospitals, and your dear authors' neighborhoods and homes. These and many other examples demonstrate how a maintenance mindset can lead to cultural and emotional well-being, and, yes, even economic prosperity.

To get a clear look at the future, we need to gain a clear understanding of the past. So let's review the history of innovation, with a basic question as our starting point: How did our era become an era of innovation-speak?

CHAPTER TWO

Turning Anxiety into a Product

A BRIEF HISTORY OF INNOVATION-SPEAK

Humanity has not always cherished innovation, or even progress. Unlike our culture today, many societies have actively opposed it. The Chinese philosopher Confucius was an innovator by almost any definition: He created a body of thought that diffused widely and had a profound influence on society that continues to this day. Yet Confucius framed his work not as a new invention stemming from his authorial genius, but as a codification of traditional ideas and values extending back more than a thousand years. Similarly, when European researchers observed the customs of Australian Aborigines, they realized that although the beliefs of those cultures had changed significantly over time, the native people themselves tended to emphasize the things that had always been as they were.

For most of recorded history, things were not much different in the West. We, like the Greek philosopher Plato, constantly hark back to our forebears, poets, and sages when searching for wisdom. Throughout Christendom, leaders stressed how their beliefs and decisions stemmed from Jesus's teachings rather than their own imaginations. In fact, during the Middle Ages, innovation—from the Latin word *innovare,* meaning "to make new"—was a distinctly bad thing. Church dogma was society's guide, and innovation, or the act of introducing new, heterodox ideas, was a heresy that got lots of people killed. *A Proclamation Against Those That Doeth Innovate* was issued by King Edward VI of England in 1548 to forbid individuals from introducing new religious rites and ceremonies. Movements from the Renaissance to the American Revolution to the German philosophers of the nineteenth century idealized ancient Greek philosophers and politicians. When we consider the broad and deep history of human thought, our contemporary obsession with the new looks downright odd.

In examining how this happened, we find it helpful to remember the distinction between actual innovation—the process of introducing new things to society—and what we call innovation-speak: the way people have come to think and talk about technological change over the past half century. It includes the word "innovation" and its variants (disruptive innovation, social innovation, impact innovation), along with a slew of Silicon Valley–approved jargon that business leaders and marketers have incorporated since the 1990s—change agent, thought leader, disruption, angel investor, intrapreneurship, design thinking, ideation, STEM education, unicorn, incubator, start-up, regional innovation hub, innovation ecosystem, innovation district—not to mention the habit of calling companies "lean" and "agile." Innovation-speak is an ideology with a set of values. Its users assume that social progress comes from the introduction of new things—even if there is short-term harm—and that change can be cultivated through a certain set of known techniques, whether they're being used by an individual or an organization. But there's no reason to believe that innovation-speak increases actual innovation in any way, and it is often unhinged from reality.

Distinguishing between actual innovation and innovation-speak can liberate us from the false promises of the latter, while preserving the genuine contributions the former provides. Actual innovation under technology-based capitalism flourished for several hundred years before the rise of innovation-speak in the late twentieth century.

Indeed, it is likely that innovation-speak began to reach its greatest heights just as actual innovation was tapering off.

BEFORE "INNOVATION"

Perhaps the most important factor that led to a positive revaluation of the notion of "innovation" was the role that actual innovation played in the massive economic, technological, and cultural changes that took place after the Industrial Revolution. The scale and scope of these changes, which began in eighteenth-century England before spreading across the world, are hard to comprehend. Some writers understandably call them a "miracle," though that miracle has always exacted significant costs, including harm to workers and the natural environment.

By the early to mid-nineteenth century, new technologies, such as steam-driven textile looms and railroads, were playing a clear role in economic and social change. The steel industry, which had a mutually reinforcing relationship with the railroads, began transforming everything from building heights to vehicle construction. Mass production would reach its apotheosis in Henry Ford's assembly lines, with their specialized machine tools and armies of workers who performed small and repetitive tasks. Automation, robots, and other technologies were added to the production process along the way, but the logic has remained the same. Mass production—or production for the masses—has vastly increased individuals' access to goods and driven down prices, which is why your closets are likely overflowing with cheap crap that would have been economically out of reach for most people a century ago.

Changes in values and social status were an essential part of this overall shift. In the eighteenth century, inventors were disparaged as "projectors," an archaic term for promoters or hucksters who pushed dubious new business ventures.[1] Very few people aspired to be inventors in a world that extolled military heroes, statesmen, and members of the nobility.

But slowly, starting in the nineteenth century, these ideals underwent a profound transfiguration. Technical creators rose in social status, particularly in England and the United States, whose leaders linked national power to the "industriousness" of their citizens.[2] Books like Samuel Smiles's *Lives of the Engineers* (1862) held up inventors and entrepreneurs as authors of industrial capitalism. And by the last decades of the century, cults of celebrity had surrounded inventors such as Thomas Edison and Alexander Graham Bell. Publications like *Popular Science* and *Popular Mechanics* spread exciting news about new technologies. The children of the twentieth century would grow up aspiring to be inventors, engineers, and creators.

These celebrations of inventors often overlooked the fact that technologies came from many heads and hands rather than a single mastermind. Eli Whitney was lauded for "inventing" the cotton gin, for instance, even though the basic technology had existed for hundreds of years in other parts of the world.[3] But observers preferred stories of individual genius to the more complex reality, a preference that continues to this day. The ordinary people who "kept the lights on" while the great men plugged away have been written out of technology's hagiographies.[4]

Edison—widely celebrated as the inventor of the lightbulb, among many other things—is a good example. Edison did not toil alone in his Menlo Park laboratory; rather, he employed a staff of several dozen men who worked as machinists, ran experiments, researched patents, sketched designs, and kept careful records in notebooks. Teams of Irish and African American servants maintained their homes and boardinghouses. Menlo Park also had a boardinghouse for the workers, where Mrs. Sarah Jordan, her daughter Ida, and a domestic servant named Kate Williams cooked for the inventors and provided a clean and comfortable dwelling. But you won't see any of those people in the iconic images of Edison posing with his lightbulb.[5]

Ironically, the cult of the inventor reached full steam only as capitalism was, in a sense, moving beyond that stage. In the early twentieth century, large corporations like DuPont, General Motors, and AT&T erected industrial research and development labs, where teams of engineers and scientists worked on difficult technical problems and developed new products. Companies did this in part to free themselves from relying on independent inventors, whose patents enabled them to ask for great sums of money. R&D labs pulled the process of invention inside corporate walls, where companies could more easily control it. During this same period, their marketing departments perfected the art of dramatizing novelty to the public, through annual model changes, advertisements, films, and auto shows. General Electric, for example, erected its House of Magic at the 1933 Chicago World's Fair, where it demonstrated cutting-edge technologies like the electric microwave.

These corporate pageants often made their case by presenting fantastical images of a much-improved future. General Motors' 1956 "industrial musical" *Design for Dreaming* (which you really should look up and watch online) is a good example. The film portrays a consumer wonderland, featuring several models of futuristic cars, endless supplies of haute couture clothing, and Frigidaire's "Kitchen of the Future," equipped with automatic technologies that would make women's work easier. "No need for the bride to feel tragic," claims one of the film's narrators, when cooking "is pushbutton magic." At the end of the film, a sleek, self-driving car ferries a husband and wife back to their suburban home, as they sing, "Tomorrow, tomorrow, our dreams will come true. Together, together, we'll make the world new."

Even by the mid-nineteenth century, technological change became tightly coupled with notions of societal progress. To some degree, the coupling is understandable. These were the years when infant mortality decreased thanks to advancements in medicine. Life spans expanded, and so did our ability to manage pain. Later on, creature comforts—electric lights, air-conditioning, soft synthetic mattresses, television, and electronic entertainments—vastly multiplied, making people's lives more convenient.

There has been a general decline in human suffering, though we should recognize that the benefits of modernity have never been equally distributed, and there are parts of the world where they are virtually unknown. Where benefits do exist, they have come with costs. Americans have grown fat, sedentary, and diabetic. Some argue that today's bureaucratic, cognitively focused, surveillance-filled forms of labor create their own kinds of suffering. Moreover, some of the biggest social advancements of the last two hundred years—such as the abolition of slavery or the expansion of voting rights to women and minorities—had little to do with technological change.

It's easy to miss this difference between technological progress and social progress, and instead see them as one and the same. In 1959, the United States opened the American National Exhibition at Sokolniki Park in Moscow. The exhibition was part of a cultural exchange program officially meant to narrow the gap between the United States and the Soviet Union, but it also served as a propaganda tool to demonstrate the superiority of American capitalism. Organizers filled the exhibition with consumer marvels like color televisions and flashy home appliances—representing the kind of home that an average American worker could afford.

On July 24, 1959, Vice President Richard Nixon led Soviet premier Nikita Khrushchev on a tour of the exhibition. The two engaged in a series of tense exchanges that later became known as the "Kitchen Debate" because it took place, in part, in the model home's kitchen. Khrushchev asked defensively if, after three hundred years of history, consumer gadgets were the best the United States could do. He predicted (quite wrongly) that the Soviet Union, which had only been around for thirty-seven years, would surpass the United States in development within seven years. Despite their differences, both Nixon and Khrushchev assumed that technological bounty could be taken as a basis for judging the merits of a country's economic system.

REBRANDING PROGRESS AS "INNOVATION"

Use of the word "innovation" took off in the years following World War II. This change in language had several causes, and professional economists—those vanguards of the dismal science—played a crucial role.

By the late 1950s, economists working in the United States faced a mystery.[6] Traditionally, economists had used changes in capital and labor to explain economic growth, but these factors simply could not explain the postwar economy's great bounty. In a 1957 article, the economist Robert Solow hypothesized that technological change was, in fact, making workers more productive and, in turn, improving people's quality of life. Productivity was profoundly affected by the introduction of new tools: Solow estimated that output per man-hour doubled between 1900 and 1940, and one of his peers calculated that roughly 90 percent of this increase came from improvements in technology. Within a decade, Solow's view that "technical progress" drove economic growth became orthodox and started an avalanche of scholarship. Today, Solow's article has been cited more than seventeen thousand times.[7]

If Solow was right, and technological change, or "innovation," as it was increasingly called, was the key to economic growth and the health of businesses, then it made sense to try to perpetuate it. The phrase "technological innovation" took off in the 1960s as economists, business leaders, policy makers, and others sought to apply its secrets to their pursuits. Government agencies, such as the Department of Commerce and the National Science Foundation, held conferences and released reports on the topic. For example, the Department of Commerce published *Technological Innovation: Its Environment and Management* in 1967. Robert Charpie, the president of Union Carbide Electronics, led the effort with about thirty other influential individuals from business, academia, and government. They put "invention and innovation" at the heart of American progress (citing Solow's 1957 article as evidence) and concluded with more than a dozen recommendations for fostering it. Yet, most of the ideas were as vague as "develop innovative and entrepreneurial talents."[8]

In many ways, the language of these 1960s documents is indistinguishable from the innovation-speak of today. Educational reformers, for example, made arguments that wouldn't have sounded out of place in Sir Ken Robinson's 2006 TED Talk, "Do Schools Kill Creativity?," which has more than fifteen million views on YouTube. "An engineering school that merely imparts information is an expensive waste," wrote the editor of *Education for Innovation*.[9] Schools, he argued, should "be a creative experience, stimulating the imagination of students and helping them to prepare themselves for the unresolved contests . . . of an imperfect world." He failed to provide any evidence that any engineering program ever "merely impart[ed] information," yet he insisted that "creative individuals are oppressed by this regime, and the real world of invention and innovation is foreign to it."[10] We'll see that creating illusory crises is a common habit among innovation-speakers.

Unlike their counterparts today, reports on innovation from the 1960s drip with optimism. At the time, the United States was on top of the world, its economy booming. Leland Haworth, the director of the National Science Foundation, told participants at a 1966 conference, "We know the key to our present industrial supremacy. From the experience of less fortunate countries we have learned that an inadequate rate of technological progress in the face of a mounting population will simply force a greater number of people to live under unsatisfactory standards."[11] Just as in the Kitchen Debate, the United States had the innovation juice. Others didn't.

The shine wore off soon, however. Beginning around 1973, the United States and many other rich nations hit the economic skids. Multiple factors spawned this downturn, and economists still debate the causes. Energy prices raised by the 1973 OPEC oil embargo and other events burdened production and consumption. Fraud and other corrupt practices triggered bank closures that rattled the financial sector. Economists scratched their heads at a seemingly impossible phenomenon: Economic growth remained stagnant while inflation rose dramatically.

Economic stagnation fit within a more general sense of American decline in the 1970s. The nation's soul had been torn out by the war in

Vietnam. Smog and litter blanketed its cities. Ralph Nader and other activists exposed the destructive, immoral practices of the corporations that had led postwar growth. Assassins killed Martin Luther King, Jr., and Robert Kennedy, and the progress of the civil rights movement seemed to be grinding to a halt. Nixon, once the symbol of postwar American capitalism with his Kitchen Debate, later became an equally appropriate icon of its decline when journalists uncovered the Watergate scandal.

In the midst of all of this upheaval, use of the word "progress" dropped significantly, and the notion of "innovation" became a kind of substitute, one that offered the ideal of change without the agony of mandating reforms in the structure of American society. As the historian Jill Lepore puts it, "Replacing 'progress' with 'innovation' skirts the question of whether a novelty is an improvement: the world may not be getting better and better but our devices are getting newer and newer."[12] Progress had always included some kind of social betterment, however; from the Progressive Era of the 1890s through the 1970s, it meant using government and policy to improve citizens' situations. Advocates for innovation, in contrast, often acted as if technological change—and the new industries that went along with it—would (at least indirectly) produce the necessary social goods on its own.

So if American society in the 1970s couldn't live up to its grandest ideals of progress—namely, liberty and justice for all—perhaps it could advance its own greatness through technology. This narrative made sense in some circles, and it is still the animating spirit behind many of our policies, such as the astonishingly low tax rates for profitable corporations like Apple and Amazon. But economists, sociologists, and historians (among others) noticed that this story doesn't add up. For one thing, by the mid-1990s, observers noted that economic inequality had increased during this period of supposedly rapid technological change—not everyone had shared in innovation's bounty.[13]

But another line of thought questioned whether technology was actually changing all that rapidly. First in the late 1990s, and then in his 2016 book *The Rise and Fall of American Growth,* the economist Robert J. Gordon argued that the most impressive technological improvements since the 1970s have been in computing, cellphones, and other digital platforms, many of which focus on entertainment. Many individuals are massively wowed by such gadgets. But Gordon argues that they do not hold a candle to the technological advances between 1870 and 1970, such as electricity, urban sanitation (clean water and sewerage), pharmaceuticals and chemicals (plastics), modern construction materials (concrete and steel), transportation (automobiles and airplanes), and computers, electronics, and modern communication systems. Other economists have argued that the start-ups of today are just playing out technologies created before the 1970s rather than creating fundamentally new ones.

There's a healthy debate over Gordon's thesis and the extent to which digital technologies have contributed to economic productivity. It's still unclear who is right, and no doubt the discussion will continue. Our point is this: By the late 1970s, economists, policy makers, and others had grown seriously worried about productivity, and these anxieties have been crucial ever since to how we talk about innovation.

As these dynamics simmered during the postwar decades, the apostles of innovation found inspiration in a region that appeared to embody all of their hopes while also bucking all of the negative social and economic trends that plagued the rest of the country. The place was called Silicon Valley, and the history of its remarkable transformation provides some lessons about technological change, innovation, mythmaking, and hype.

THE RISE OF THE INNOVATION EXPERTS

To the extent that there was a recipe for the rise of Silicon Valley, the most important ingredients were prepared by Frederick Terman, dean of Stanford's School of Engineering from 1944 to 1958 and university provost from 1955 to 1965. Terman redefined Stanford by aligning the school's research with military priorities, and using money from defense contracts to recruit faculty members from the growing electronics industry. Terman also encouraged Stanford faculty to work as consultants in the private sector and create their own companies. The resulting flows of knowledge, people, and technology between the military, industry, and university earned Terman accolades as a father of Silicon Valley.

The success of Terman's model was signified by a steady parade of start-up companies and blockbuster technologies: powerful and ever-cheaper microprocessors made by Intel; the defining protocols of the Internet (which a Stanford computer scientist helped create); user-friendly computers and integrated media devices made by Apple; and platforms for sharing text, audio, and video made by Google and Facebook. Their technologies, in combination, made it possible to watch live videos from outer space on a device that fits in the palm of your hand. We are only beginning to reckon with the far-reaching consequences of the technologies that Silicon Valley built.

Just as the Silicon Valley region was admired as a model of success, so were its leading companies. Apple, Google, and Facebook built lush corporate campuses, with open workplaces sporting beanbag chairs and ping-pong tables, and billion-dollar profits were the norm. More than a business strategy, the ideology of Silicon Valley encompassed an entire culture that was supposedly hipper, kinder, and more open-minded than the "organization men" who ran the bureaucratic giants of yore. Clothing was casual: black T-shirts, jeans, and the ubiquitous hoodie.

The area generated wealth along with its dazzling technologies—a fact that did not escape politicians. Cities and regions around the world embarked on quests to capture the magic for themselves: Silicon Alley in New York City, Silicon Prairie in the Midwest, the Silicon Valley of India (aka Bangalore), Chilecon Valley (Santiago), and Silicon Wadi (Tel Aviv).[14] All of these efforts, and dozens more, played with the basic ingredients in Terman's recipe—connecting universities, investors, and entrepreneurs in a clearly defined region—to reproduce the special sauce.

Silicon Valley's innovation-speak didn't always age well, and its breakthrough concepts—disruptive innovation, hackathon, angel investors, thought leaders, intrapreneurs, bodystorming, tech meetup, gamification, Big Data, AI, change agents, ideation, start-ups, incubators, empathy, Design Thinking, thinking outside the box, unicorns—eventually became tired clichés. Even so, the Silicon Valley mystique helped spread the gospel of innovation-speak into all areas of twenty-first-century society, as all kinds of firms tried to model themselves on start-ups. Throughout the United States and beyond, schools bought iPads, medical insurance companies released apps, museums and libraries innovated, government agencies gamified public services, and elected officials wouldn't stop tweeting. Organizations across the globe—including old, staid firms like General Electric—tried to tap into the mojo of Silicon Valley start-ups, with varying degrees of success.

As innovation became a priority, it created demand for what the historian Matt Wisnioski calls innovation experts—a new breed of individuals, often consultants, who offered up visions and plans for how to make individuals, organizations, even cities, regions, and entire nations more innovative. And by "offered up," we mean *sold:* You could make good money if you came with an enticing theory of innovation.

The late Clayton Christensen, a long-time professor at Harvard Business School, is a good example of this phenomenon. In his 1997 book *The Innovator's Dilemma* and a stream of subsequent publications, Christensen spelled out the idea of "disruptive innovation," a process by which a new technology or business model can upend existing markets, firms, or products. Christensen's concept caught on like a new, great, powerful drug. All around the world, corporate boardrooms filled with smoke as executives took turns either imagining themselves as the next disrupters, creating the killer app that would blow away the competition, or worrying that they themselves would be disrupted by some unseen start-up in their field.

Those who were really concerned could hire Christensen's consulting firm, Innosight, which would help them see all of the opportunities for disruption looming large. Christensen wasn't the only person to push the idea (or profit from it). For example, in a post titled "Top 10 Rules of Entrepreneurship," Reid Hoffman, one of the founders of LinkedIn, listed "Look for Disruptive Change" as his first rule. Outlets like GE Reports, a division of General Electric, published pieces like "How to Create Disruptive Innovation." The jargon caught on, and then dribbled out to the rest of us through business magazines and TED Talks.

There is a problem, however: Both popular and academic publications have called into question the notion of disruptive innovation since its inception. For example, a central case study of *The Innovator's Dilemma* focused on young, upstart firms that "disrupted" the disk-drive industry in the 1980s. But the firms that were still around in 2014 were the companies that "led the market in the nineteen-eighties," Jill Lepore pointed out in *The New Yorker*.[15] "In the longer term," she concludes, "victory in the disk-drive industry appears to have gone to the manufacturers that were good at incremental improvements, whether or not they were the first to market the disruptive new format."

Moreover, even though Christensen and other Silicon Valley thought leaders have encouraged companies to aim for disruption, there is zero evidence that doing so has led to more new products or business models or upended existing technologies or industries. Disruption is not something you can work toward or plan. For example, neither the Department of Defense engineers who created the underlying protocols for the Internet, nor Tim Berners-Lee, the inventor of the World Wide Web, meant to shake up entire industries, from journalism to home entertainment, retail, and travel planning. Yet that is what they did. Their creations, which they achieved via incremental improvement, had far-reaching and unplanned consequences. Actual innovation proceeds through small steps, not grand strategy.

Another example of a hot idea that unraveled on closer inspection was the notion of a "Creative Class" promulgated by the urban planner Richard Florida. In *The Rise of the Creative Class* and subsequent publications, Florida put forward a kind of *Field of Dreams* "build it and they will come" theory of public planning, only for hipsters instead of baseball players. Florida argued that the presence of the "Creative Class," including "scientists and engineers, university professors, poets and novelists, artists, entertainers, actors, designers, and architects," led to a virtuous cycle of investment and economic growth.[16] To tap into the power of this class, cities needed to foster the kinds of urban features—cool bars, art galleries, high-end coffee shops, and bike-sharing programs—that appealed to them. If cities wondered how they should go about doing this, they could hire Florida's consultancy, the Creative Class Group, which describes itself as "a boutique advisory services firm comprised of leading researchers, communications specialists, and business advisors."

Study after study has failed to support Florida's arguments.[17] For instance, Florida claimed that the migration of creatives to cities was driving growth, but many other studies found the opposite causation: The primary reason workers moved was because *jobs* were available. The sad reality is that lots of places tried out Florida's ideas and found they just didn't work. As the journalist Frank Bures puts it, "Millions of dollars had been spent by hard luck towns across the Rust Belt in the hopes that a coffee shop, a bike path and a co-working space would restore their postwar industrial glory. Yet for cities that took Florida's theory to heart, like Youngstown, Cleveland, or Duluth, the boom never really came."

Who should be trusted as an "expert"? Stephen Turner, a philosopher and sociologist, argues that we should distinguish between people who have demonstrated efficacy and those who have not. We should trust physicists when they make claims about physics, for example, because they can do stuff with their knowledge like make bridges that work or rockets that fly. But nobody on this planet *knows for certain* how to make more innovation in general, and if someone claims to, he or she probably has something to sell you.

For the purposes of this book, however, another aspect of *The Rise of the Creative Class* stands out. Florida didn't just argue that cities should pour all of their attention into keeping innovators comfortable, happy, and entertained. He actively denigrated what he called the "Service Class," people working in non-innovative professions like food service, hospitality, and personal care. As Florida wrote, many members of the Service Class "have no way out [and] are stuck for life in menial jobs," living a "grueling struggle for existence amid the wealth of others."[18]

What would Florida's Creative Class proposals do for these ordinary people who kept the world running? Not much. In fact, they may even have done some harm. Torn between the haves and have-nots, cities were becoming increasingly unequal places, and Creative Class policies may have exacerbated this divide by encouraging gentrification. In 2017, Florida published *The New Urban Crisis: How Our Cities Are Increasing Inequality, Deepening Segregation, and Failing the Middle Class—and What We Can Do about It.* Many saw the book as Florida's mea culpa for getting it wrong.[19] (Though in some ways, he's doubled down on the Creative Class theory in years since.)

Another innovation product that's recently come under scrutiny is called Design Thinking. Its roots extend back to the 1950s and '60s, and it began as a reasonable discussion within the professional field of design. The most popular form of Design Thinking today is associated with the fabled design firm IDEO, which is most famous for creating the original Apple mouse in 1980. David Kelley, one of the company's founders, asserted that a core aspect of good design involved "empathy," which he characterized as "the ability to see an experience through another person's eyes, to recognize why people do what they do."[20] Many saw Kelley's notion of empathy as an important intervention in design and engineering, which had increasingly come to rely on quantitative methods for creating things and sometimes seemed to forget that users existed at all, leading to the creation of clunky, unusable junk that no one wanted.

Kelley and others codified Design Thinking into a curriculum that taught "empathy" and other techniques they took to be fundamental. Kelley was a famous and influential figure—a kind of guru—

and Design Thinking had some powers and advantages. Colleagues who teach design in universities tell us that it is a handy way to introduce students to some fundamentals of the discipline.

But proponents quickly overhyped the method by selling it as a technique that could solve problems in any and every domain of human life. That shift resulted from changes at IDEO itself. As the communications scholar Lilly Irani has described, by the middle of the first decade of the twenty-first century, IDEO faced increasing competition from Chinese manufacturers and design firms. In this competitive environment, Irani writes, IDEO "was no longer able to command high rates for engineering-oriented projects."[21] The company decided to, as Irani puts it, "de-emphasize the design of things" and move instead into consulting, imitating older outfits like McKinsey. More and more, Design Thinking became IDEO's central output, first as a form of business consulting and corporate education and later as an educational product that could be sold to the masses. When Irani interviewed machine shop workers at the firm, one of them put it like this: "There's been a shift to less mechanical and to more mystical." And part of that mysticism involved imbuing Design Thinking with the seemingly magical ability to *make everything better*.

Around this time, Kelley approached a rich fan-client about "creating a home for Design Thinking" where the method could be taught to individuals from a wide variety of backgrounds and industries.[22] A $35 million donation founded the Hasso Plattner Institute of Design—also known as the d.school—at Stanford. Stanford began charging individuals nearly $15,000 for a four-day Design Thinking Bootcamp called "From Insights to Innovation." Or one could pay IDEO $399 for a self-paced, video-based Design Thinking course, "Insights for Innovation." If you think those titles are awfully close, you're not alone. The line between the d.school and the design firm wasn't always clear.

But the more Design Thinking became sold as a panacea, the more it became watered down and the more its limits became clear. In her viral talk "Design Thinking Is Bullshit," the designer Natasha

Jen, a partner at the design firm Pentagram, lists example after example where Design Thinking was used to produce something that could have been reached through any other method, including common sense. She shows how designers painted cartoons on the wall of an MRI machine for children to make the kids feel more comfortable. She then clicks over to products made by Oil of Olay and IBM through Design Thinking that look exactly like . . . products made by everyone else in those companies' fields. There's nothing truly novel here. In Jen's view, "Design Thinking packages a designer's way of working for a non-designer audience . . . claiming that it can be applied by *anyone* to any problem."[23] It's a product you can sell to individuals and firms who are desperate to be innovators but do not want to go through years of training to become actual designers.

We see important similarities between the Christensen, Florida, and Design Thinking stories—most of all, that they center on *consulting*. An enormous market of organizations and individuals yearn to be innovators and will pay big bucks to become one. Innovation experts benefit from a deep-seated human frailty, which the philosopher Ludwig Wittgenstein called "the craving for generality." General statements are crucial for living, of course. You aren't going to last long if you can't learn principles like "fire burns" or "that red berry is poisonous and will kill you dead." But Wittgenstein's point is that we often hanker for universalities in cases where the complex world evades such easy summary. If we consider the broad panoply of things we call innovations—new things entering the world—we will quickly see there is no common pattern for how they come into being and spread through society.

What's striking when we look back at the history of technology and business is how many successes did not fit anything remotely resembling the Design Thinking process. Henry "Any Color, So Long as It Is Black" Ford had no love or respect for his users, yet his products did more than okay. Steve Jobs assumed that customers did not know what they wanted but rather had to be shown by geniuses like . . . Steve Jobs. Moreover, many crucial innovations *create* needs rather than respond to needs that already exist. Many homeowners saw no need for electricity when it was first introduced on the market around 1900. They had to be convinced that it was useful. Most innovation and subsequent development is unpredictable and cannot be planned, ideated, or prototyped. No one working on commercializing the Internet in the early 1990s foresaw the emergence of meme culture or Instagram "influencers."

When we take all of these examples together, it's easy to see there is no recipe for creating innovation. Yet, as a friend put it to us, "Well, you can *sell* recipes." *If only* a recipe like Design Thinking could produce "reliably innovative results in any field," the world would be a much simpler, easier, and more manageable place.[24]

We are not opposed to actual innovation. In fact, the two of us have published articles about how to improve innovation policy in the United States. Our point is that we should resist the notion that anyone on this planet knows how to increase the rate and quality of innovation in general, and we should all be skeptical of anyone who makes such claims. The late economist Nathan Rosenberg and others who have written deep studies of innovation have tended to emphasize incremental changes and long processes of continual improvement. Indeed, most innovation and most of the changes that have contributed to the massive transformations of the last three hundred years are of this sort. The incrementalist vision suggests that the best advice one can give about innovation is this: Take care, pay attention, and *do your job*. It's not the kind of message that will attract multimillion-dollar endowments to universities or enable your dear authors to open up a consultancy and get filthy rich. But we believe it's a more honest picture of how technological change actually works.

To move beyond innovation-speak, we need to embrace different, more nuanced ways of thinking about technology. (We stress different, not *new,* ways of thinking. Improvement will involve drawing on long-standing traditions of thought as much as anything else.) The next step down this path is to decouple "innovation" from "technology," and to think hard about what we want from technology in the first place.

CHAPTER THREE

Technology after Innovation

WHY MAINTENANCE IS INEVITABLE—AND USUALLY NEGLECTED

We'd like you to indulge in a thought experiment with us. It has three steps.

Step 1: Take a moment, look around you, and make some mental notes. What technologies do you see? What technologies in your vicinity are new, and which ones are old?

Go ahead. We'll wait.

What did you see? If you're inside a building, you probably saw walls—layered combinations of steel, cement, wood, screws and nails, and paint. You also saw furniture, electric lights, windows, carpets, a sink, or maybe—if you're *really* enjoying this book—a toilet. Of the newer technologies that surround you, one of them would be carpet, which is typically made from synthetic materials invented in the 1930s (nylon) and 1950s (polypropylene). Another would be LED lightbulbs, which were invented in the 1960s and brought to the mass market in the last decade. The newest technologies you saw were likely products of the digital age: televisions, computers, or perhaps a voice-activated surveillance device like Amazon Echo or Google Home.

Step 2: Which technologies are essential for your well-being, and which ones could you do without? Let's say you had to banish one of these things from your home: Amazon Echo or reinforced concrete. Which one would you choose? Let's say you need to banish one of these things from your local elementary school: glass windows or iPads. Which one would you choose?

Step 3: Recall the various technologies you used in the past twenty-four to forty-eight hours: What was old, and what was new? Surely there were some digital devices and apps in the mix, but what about the other things? When you think about technologies, do you consider the car, bike, bus, or train that brought you to work or school; the appliances that cooked your food; the soaps you used to clean your body and clothing; the infrastructure that delivered gas, electricity, and water to your residence or workplace?

Our goal with this thought experiment is to draw your attention to how widespread ordinary technologies are in your everyday life. Their importance stands in stark contrast to our culture's obsession with innovation, the phenomenon we documented in chapter 2. A heightened awareness of how we actually use technologies is an essential first step for combating the deeper problems we investigate throughout the book, including our obsession with innovation and the corresponding failures to invest in the ordinary technologies and people who keep our world running.

In this chapter, we'll reintroduce the notion of maintenance, with particular attention to its central paradox: It is, simultaneously, both absolutely necessary and usually neglected. This is a cruel irony, since maintenance is the key to ensuring that the benefits of technology are felt in their full depth and breadth. Think again about the technologies you depend on—and then think about what happens when they aren't maintained. Rooms go dark, toilets clog, windows leak, cars break down, bridges collapse.

Maintenance preserves order. It is the constant war against entropy—the second law of thermodynamics, which states that, over time and without intervention, every system will decline into disorder and randomness. Although you wouldn't know it from histories that fixate on innovation and inventors, much of human history is, in fact, stories of stability: of how societies coordinate labor to maintain the large-scale public systems we've relied on since ancient times.

To ensure that our society avoids the consequences of undervaluing maintenance, we need to recalibrate how we think about technology more generally. We'll start by considering some of the words that have slipped into our everyday actions and even our everyday vocabulary—starting with the suddenly ubiquitous term "tech."

"TECH" VS. TECHNOLOGY

The word "tech" is everywhere. Newspapers ponder the future of "Big Tech"; analysts tell us about the fortunes of "tech stocks"; and *The New York Times* has a "Tech Fix" column that covers the Uber, Google, and social media beat.[1] When you read or hear "tech," it is usually shorthand for Internet-based digital devices, services, and apps. "Tech" companies such as Google, Facebook, Microsoft, Apple, and Amazon profit from some combination of advertising, convenience, and user addiction.

But it's misleading to use "tech" as a proxy for "technology." The term is too narrow. Technology, as a human phenomenon, is much broader and deeper—it encompasses all of the materials and techniques contrived by human civilization, including non-digital technologies such as guns, sidewalks, and wheelchairs. When we reduce "technology" to "addictive digital devices and their applications," we discount thousands of years of ingenuity and effort, and needlessly focus our attention and resources on a very small band of human experience.

The definition of "technology," like that of every powerful concept, changes over time and follows the interests of the people who define it. English speakers rarely used the word "technology" before the 1930s. "The useful arts," "applied science," "machines," and "manufacturing" were more prevalent terms. One exception was a small group of new American universities that incorporated the word into their names in the mid-nineteenth century—the Massachusetts Institute of Technology and the Stevens Institute of Technology, for example. In these cases it was well understood that "technology" referred to those schools' educational mission: training mechanics and engineers.

In the late nineteenth and early twentieth centuries, these college-educated engineers banded together into professional groups, in part to advance their status within universities and corporations. As their fortunes rose, they systematically diminished the contributions of artisans and other groups of expert workers—women, immigrants, African Americans, and the uneducated poor—whose labor was also essential for the machine age. By the 1930s, "technology" was nearly synonymous with the visions of progress and material abundance that were showcased at World's Fairs in Chicago (1933) and New York (1939). And it almost always referred to the exploits of white males from the middle and upper classes.[2]

Technology, in this new view, was the driving force of history. Not only was it pointless to resist; it was also wise to embrace it. Although some critics worried about a capitulation to technology (Martin Heidegger's 1954 essay "The Question Concerning Technology" is still a foundation of university seminars on the ethics of technology), they were in the minority. The notion of technology as a progressive and inexorable force matched the everyday experiences of Americans in postwar consumer society. It also suited the corporations and universities whose business models depended on the public's acquiescence.

The idea that technology was now and forever a force of progress survived the environmental awakening of the 1970s. Inspired by the rise of new electronic devices, computer networks, and software in the 1970s, '80s, and '90s, pundits coined some new terms—"information superhighway," "cyberspace," "compunications"—to describe what they believed to be a world-historical shift to an information age. But none of these neologisms were sticky enough to characterize the amazing new world being created by the Internet, mobile telephony, and the proliferation of handy computer applications. The term "ICT," short for information and communication technologies, never really rolled off the tongue. Instead, networked digital electronics became so pervasive, and so mesmerizing, that corporations, journalists, and eventually everyone else silently and mindlessly complied in referring to these things as "technology," or just plain "tech."

This bastardization of technology concerns us because it embodies the very worst of innovation-speak—it's shallow, focused on the new and the digital, and excludes and degrades all of the other meanings of technology that are so important for our society. Technology (as an object) isn't merely digital and new; and technology (as a process) isn't merely innovation.

To find a healthier approach to technology, let's start by considering a simple definition: *Technology includes all the things humans use to help them reach their goals.* These things include tools (including ordinary objects like cutlery), buildings, cloth, streets and sidewalks, and the pipes, pumps, and wires that we use to transport water, waste, energy, and information. The late novelist Ursula K. Le Guin had an even simpler definition: "Technology is how a society copes with physical reality." In her "Rant about 'Technology,'" she noted that "we have been so desensitized by a hundred and fifty years of ceaselessly expanding technical prowess that we think nothing less complex and showy than a computer or a jet bomber deserves to be called 'technology' at all."[3]

Let us emphasize two fundamental points again. First, technology is not just "tech"; it's more than digital consumer devices and apps. Second, technology is not just innovation. Most of the technologies we rely on are so common that we barely think of them. They fade into the background of our culture and—crucially—our financial planning. And yet, it's essential that these things continue to function—as anyone who has experienced a power outage or water main break can attest.

Since technology is not just innovation, and the tools that we use daily consist mostly of old things, then a logical next step is to think about not only *what* technology is but also *when* it is.

A piece of technology passes through three basic phases: innovation, maintenance, and decay. We spent the last chapter talking about innovation; now it's time to focus on what happens *after* innovation.

When humans interact with technology, they generally do not create it; rather, they use and maintain it. We spend a great deal of time cleaning our houses, refueling and fixing our vehicles, updating our computers and apps. (Or, in many cases, we assume that other people will complete these tasks for us.) Moreover, as humans have introduced more and more new things to the world, we've created more and more things that must be maintained. Otherwise these things decay, and so do the societies that rely on them. Of course, nothing lasts forever—senescence and decay are natural and inevitable for humans, animals, and technology.[4] But there are a lot of things that will stick around for a while, if we take care of them.

At an intuitive level, everyone knows that maintenance is important. As children, we learn that maintenance is an essential component of our daily routines. Bathing, brushing our teeth, exercise, eating and drinking—all of these activities maintain the health of our bodies and, depending on what we eat and drink, keep them in good working order.

We also keep our material possessions (aka technologies) in working order through regular or semi-regular routines: vacuuming, sweeping, dusting, checking the oil and tire pressure in our cars, wiping down our kitchen counters, cutlery, and dishes, clearing out old photos and files from our computers and phones, and making sure that walkways, decks, gutters, and drainpipes work as intended. And we search for ways to maintain sanity and peace of mind, through activities such as prayer, meditation, and time for reflection or decompression. Indeed, many people find respite in activities that combine the maintenance of different things, such as yoga (body and mind) or motorcycle repair (mind and technology).

Maintenance is, obviously, crucial for individual health. But how does it affect the health of societies? One answer comes from the artist Mierle Laderman Ukeles, who posed a famous challenge to her generation: "After the revolution, who's going to pick up the garbage on Monday morning?" Ukeles became the first artist in residence for the New York Department of Sanitation in 1978, at a time when sanitation workers were striking and City Hall couldn't find the money to keep its streets and sidewalks clean. Her work challenged New Yorkers to pay attention to the things they took for granted. The hours she spent following, interviewing, and shaking the hands of sanitation workers were seen as radical acts: calling attention to the value of their lives and labor.[5]

Ukeles's work shed light on an essential fact of technology—that it requires maintenance and care. Subsequent work by scholars such as Carol Gilligan, Nel Noddings, and Virginia Held has argued that care is fundamental for all societies; it is frequently feminized (that is, powerful actors in society treat care work simply as obligations of women); it is usually undervalued and underpaid; and it appears in all social settings, from families and friendships to bureaucracies and laboratories.[6]

When things go wrong, the first place we should look is to see if the relations of care are healthy. Nancy Fraser, a political philosopher at the New School, has pointed out that the excesses of capitalism and global ecological catastrophes have a common theme: They are the costs of neglect, and the consequences of a society that values the individual accumulation of wealth above the common good. Symptoms of this crisis are evident in the feeble American healthcare system; the flourishing of Uber while public transportation systems crumble; and the mounds of postconsumer plastic that gather in our oceans. To put it another way, if we devoted more effort to caring about one another—our health, our mobility, and our environment—we would naturally devote more effort to our collective maintenance. It's difficult to see how these forms of maintenance, or *any* forms of maintenance, can be performed in the absence of care. In turn, it's difficult to see how any technological civilization could survive without it.

Caring thus refers to both the provision of necessary attention for health and welfare and a feeling of concern that rises above rational or functional relations. To value care work—as Ukeles did—is an essential step toward moving past a culture of narcissistic materialism. Yet an obsession with novelty and innovation can blind us to the value of care. And as we'll see, some of the salient characteristics of care—particularly its ubiquity and its undervaluation—are constant themes in the history of maintenance work as well.

MAINTENANCE IN HISTORY: ESSENTIAL, YET NEGLECTED

If you go to a bookstore or a library and look for histories of technology, the shelves will be filled with biographies of great inventors like Edison, Tesla, and Bell, and stories about the creation of planes, trains, and automobiles. Yet, as we have seen, most human activity centers on *using* technologies, not creating them. Stories of our everyday interactions with the material world have largely gone untold.

In the next few pages, we'd like to sketch a different approach to these stories, one that foregrounds maintenance and care, so that we can provide a clearer picture of how we arrived at our present condition. You'll see two persistent themes in our retelling of the history of technology: Maintenance and care are *essential,* yet they are regularly *neglected*.

For as long as humans have made technologies, from clothing to weapons to woven baskets, those objects have required upkeep and attention to remain usable. Ancient clothing uncovered by archaeologists, for instance, often shows signs of mending. And some cultures went so far as to codify maintenance and repair requirements in their religion and rules. Jewish books of scripture and law, for example, outline inspection and maintenance routines for holy artifacts and texts.

Cultural expectations—including senses of cleanliness, order, and duty—play an enormous role in determining what kinds of upkeep we choose to do. To explain this point to us, Pamela O. Long, a historian of medieval technology and a MacArthur Fellow, used sewage as a case study. The streets of Rome in the fifteenth and sixteenth centuries were strewn with garbage and waste from animals and humans alike. "From a modern point of view, they were perfectly revolting," Long says. She examined more than two hundred years of records and found "papal bull after papal bull, regulation after regulation" demanding that the streets be cleaned. "It was incredibly difficult to do, in part because there was really no institutional or bureaucratic structure set up to make sure it happened." Responsible for maintaining the roads were the so-called masters of streets, civic elites with roles in government, "but they did not head a department with permanent paid employees with specific jobs and equipment aimed toward trash collection and sewage disposal."

Long concludes that maintenance should not be taken for granted, even in the present day: "I would say it's not a given at all. It includes a whole group of technological systems that were not in place for much of human history. In cities, maintenance was a huge struggle that had numerous failures."

Some deeply significant trends in the history of maintenance were sparked by two transformations in the mid-nineteenth century: urbanization and the rise of industrial capitalism.

Ninety-four percent of Americans lived and worked in rural spaces in 1800, mostly on family farms. That figure dropped to 74 percent by 1870, and it was down to half by 1920.[7] (Today less than 20 percent of Americans live in rural areas.) Where were all these former farmers going? Well, of course, they were moving to cities where they could find more lucrative, though often onerous, wage labor. By the late nineteenth century, large corporations were sprouting up in cities across the nation. Many of these big companies were "capital-intensive," requiring large investments in materials and equipment before their work could even really start. Railroads, steelmakers, refineries, and factory-based mass production all fit this bill. And the technologies at the center of their businesses required constant maintenance and repair. Without upkeep, their machines failed, production halted, workers stood idle, and money went to waste.

As businesses came to use and rely on complex technologies, new occupational roles emerged. "Mechanic"—or "mechanick" in a now-obsolete spelling—was a word applied broadly to artisans and manual laborers as early as the sixteenth century.[8] Shakespeare mentions "mechanickes" in his play *Coriolanus,* for instance. But the word was used more restrictively by 1800, referring to skilled individuals who operated machines like steam engines and waterwheels.

Railroads were the focal point of mechanical societies by the end of the nineteenth century, and for good reason. They served as the infrastructure for American commerce and expansion, and the entrepreneurs and cities that learned to take advantage of rail transport likewise enjoyed its spoils: Vanderbilt, Carnegie, Rockefeller, Mellon, Frick; Chicago, Pittsburgh, Denver, New York City.

The titans of the industry knew they would profit only if their railroads were reliable pathways for American commerce, so they quickly became the testing beds of work and thinking around maintenance. Railroads employed hundreds of people and owned a staggering amount of physical assets: from engines and cars to rails and roadbeds to equipment and buildings. All of it required care and upkeep, and the experiences of railroaders charged with maintenance are still deeply relevant today.

Railroad executives often looked down on maintenance departments (or at least that's how maintainers felt), but roadmasters—the individuals who counties and cities employed to keep up roads—had a grand vision of the role railroads played in society. In an address at the first meeting of the Roadmasters' and Maintenance-of-Way Association, one roadmaster declared, "The railway is the modern highway of a modern commerce. . . . The sovereigns of empires, the chiefs and presidents of republics, with the vast multitude of their followers, intrust their lives to its mighty power, while the management of this great factor in the world's progress must carry the dread responsibility of this burden thus confided to its keeping." Railroads may have led to national economic growth and boom times in cities like Chicago and St. Louis, but without maintainers, nothing would have happened, except for a lot of gory accidents. For example, on December 28, 1879, a portion of the Tay Bridge—an iron bridge in Scotland connecting Wormit and Dundee over the Firth of Tay—collapsed into the water nearly one hundred feet below. A train fell into the sea, killing all aboard, estimated at seventy-five people. The investigation found high winds to be the primary cause, but poor maintenance practices were also a central factor. As one investigator (with the title of commissioner of wrecks) put it in a report, "Can there be any doubt that what caused the overthrow of the bridge was the pressure of the wind acting upon a structure badly built and badly maintained?"

Despite their lofty self-image, roadmasters found that railroad executives undervalued maintenance. Track work was strenuous, backbreaking labor, the lowest of the low in the hierarchy of railroad jobs. Railroads often treated these employees terribly, working them to the bone at poverty wages. In 1889, for instance, a railroad in Waycross, Georgia, paid new track workers 75 cents a day, raising to a maximum of $1.50 a day—in 2018 dollars, that's between $14 and $28 per day, or $3,640 and $7,280 per year. If a laborer was good and/or lucky, he might be appointed a foreman, making $2.25 a day. At one meeting, a roadmaster explained that he purposely paid meager wages to each worker. "You will make him so poor that he cannot get away?" another asked. He responded, "That is the idea." Moreover, men injured, maimed, or disabled on the job were often let go without compensation.[9]

Given how difficult the work was, managers had specific ideas about what made someone a good track worker. In the late nineteenth and early twentieth centuries, the roadmasters had highly racialized images about what made a worker smart and capable. Mexicans, Italians, and eastern Europeans were not "men of any intelligence." Black workers had it even harder. In *Steel Drivin' Man: John Henry; The Untold Story of an American Legend,* the historian Scott Reynolds Nelson argues that the man at the center of the famous folk song "John Henry" was likely a black convict coerced to labor as a track worker in Virginia. Many of these forced laborers faced arduous conditions that caused their deaths. Roadmasters gave their workers cocaine to ease the toil and numb the pain. "For road bosses, addiction may have been an added inducement, for it kept men from leaving the gang," writes Nelson. "For black men's health, cocaine was terrible."[10]

These patterns in the railroad industry—where maintenance work was both very important yet neglected—were present in other leading industries of the era, including steel, manufacturing, chemical plants and refineries, and electrical and telephone systems. While it may be hard to imagine now, wealthy individuals often set up their own electrical, telephone, streetcar, and other systems during this period. Self-styled experts published a number of how-to books explaining what such systems required—a bit like today's business self-help books. Invention-centered histories have ignored the proliferation of how-to volumes that consistently emphasized the systems' need for maintenance and repair.

As American industrial production flourished in the twentieth century, one rapidly developing market was the creation of technologies for the home—a major element in the rise of consumer culture. Consumers bought electric phonographs, radios, fans, toasters, refrigerators, vacuums, and washing machines. Automobiles increasingly replaced carriages in family driveways. With all of these technologies, industrial notions of maintenance and reliability collided with long-established traditions and tensions of maintenance and care in domestic spaces.

Many of these technologies were meant to save homeowners time and labor, particularly when it came to maintaining the home or "keeping house." More to the point, corporations marketed devices like vacuums and washing machines with promises that they would save housewives' effort. These claims were nonsense, as the historian Ruth Schwartz Cowan demonstrated in her classic *More Work for Mother*. While the new technologies spared women from the more physically demanding side of tasks—like the more grueling aspects of laundry—they led to higher standards of cleanliness and women doing more housework, work that was by definition never finished.

Consumers were increasingly forced to rely on outside experts to keep their technologies in working order. Leaving aside potentially deadly electrical systems and appliances, even basic plumbing exceeded many homeowners' know-how, or at least their patience. The term "repairman," virtually unheard of before 1850, burst on the scene in the first years of the twentieth century. Repairmen did many tasks for many different kinds of groups, but their primary focus was always on servicing consumer-owned technologies. Over the course of the twentieth century, repair entrepreneurs opened up waves of shops to service new consumer gadgets—first radios, then televisions, then computers—a phenomenon that continues down to today's iPhone repair stores.

Perhaps the most iconic place of maintenance and repair in the United States is the auto mechanic's shop. Working as an auto mechanic appeared to be a promising strategy for men to climb into the middle class. Good jobs could be had maintaining the automobiles of rich owners—the only people who *could* afford cars at first. Over the course of the twentieth century, however, auto work came to be seen as a "dead-end job" that only unpromising students were encouraged to enter. The status of auto maintenance and repair also suffers because mechanics make consumers anxious: Car owners depend on them and are vulnerable to their expert knowledge. Consequently, the worries and rumors about auto mechanics who rip off their customers are a constant theme in the history of auto maintenance.[11]

Automakers responded to these concerns by trying to make cars simpler to maintain. The earliest automobiles required constant maintenance and repair. Drivers in the 1910s needed to have some technical know-how. But car manufacturers like General Motors invested heavily in making auto engines and other systems more reliable—not only to relieve customers of the hassles of doing their own maintenance but also to meet other important objectives like safety. Makers of other technologies followed suit and simplified user maintenance regimens, though there are many well-known exceptions. (One appliance repairman griped to us about how refrigerator manufacturers added flashy features like ice makers to wow and attract consumers; the only problem was these extras regularly failed, which meant angry owners needed to call him again and again and again.)

Companies also developed warranty programs, promising that consumers would not be left with a broken product. Maytag eventually riffed on this idea when it gave birth to its Maytag Repairman ad campaign: The Maytag Man is "the loneliest guy in town," who takes up solitaire and crossword puzzles because Maytag's products are so good no one ever calls him to make a repair.

Such is the legacy of maintenance in the industrial age: We understand it's important, but if we had our way, we would never have to worry about it.

THE RISE OF THE MAINTENANCE EXPERTS

Two important concepts emerged early in the twentieth century: deferred maintenance and preventive maintenance.

With the emergence of capital-intensive enterprises like railroads and factories, engineers and accountants needed a way to record how machines and physical assets degraded in quality as they aged, especially if they were not properly cared for. By the 1890s, book publishers were printing depreciation tables and works like Ewing Matheson's page-turner, *The Depreciation of Factories, Mines, and Industrial Undertakings and Their Valuation*.[12] Around that time, deferred maintenance was often used to describe a reserve account where organizations and governments set aside money to deal with maintenance that had been put off. Sometimes this is still true today, but since the 1910s, "deferred maintenance" has more often been used to track work left undone, with no plans for fixing it and little idea where the cash will come from.

The notion of preventive maintenance arose in the 1920s and '30s. It arose from a grudging acceptance that maintenance was necessary, but that it should be performed in a planned, orderly way that did not disrupt production. As the journal *Maintenance Engineering in Plants,*

Mills and Factories explained in 1931, "Formerly maintenance was thought of as repair. Today industry considers repair as a minor phase of maintenance," which also involved systematic inspection of all aspects of a building. The dream was that prevention would ward off all failures, malfunctions, and accidents before they ground other work to a halt, though this impossible dream never became perfect reality.

After World War II, the notion of preventive maintenance traveled from the private to public sector. Programs like Eisenhower's Interstate Highway System built large, economically and socially significant infrastructure. While this new infrastructure aided economic growth and improved some individuals' quality of life, it also encumbered local governments and citizens with mounting maintenance costs. In 1954, two years before construction on the Interstate Highway System started, annual maintenance costs for state-administered highways were about $648 million. By 1974, those costs had ballooned to $2.7 billion a year. Some of these costs stemmed from runaway inflation and other economic problems, but they were also simply the products of an unwillingness to account for the future costs of new construction. Predictions were often way off. In 1968, the American Association of State Highway Officials estimated that highway maintenance costs would reach $2.5 billion in 1977. The nation hit the threshold in almost half the predicted time, by 1973.[13]

Governments responded to these costs by deferring maintenance. This was a clear step backward in terms of reliability. Putting off the work resulted in degraded systems, of course, but it could also lead to accidents and other threats to public health. Beyond the physical harm caused, expensive lawsuits were also filed against state and local governments. In one example from the 1970s, a state failed to properly mow the low-maintenance crimson clover it had planted in a highway median. The clover eventually grew tall enough to block headlights on the road, which led to a gruesome traffic accident that killed a young girl. Courts held the state's Department of Highways responsible for the crash. All over the country highway departments faced litigation relating to the maintenance of roads and surrounding vegetation.[14]

By midcentury, planners and engineers were motivated to find new ways to prevent failures and breakdowns. Through new approaches for "predictive maintenance," they developed techniques and tools to keep industrial machinery reliable and thereby preserve the good working order of industrial society.

The roots of predictive maintenance lie in T. C. Rathbone's 1939 paper "Vibrational Tolerance," in which he asserted that machines vibrated more as their conditions deteriorated. If engineers and managers could measure vibrations, he reasoned, they'd be more likely to spot problems before machines broke and halted production. Several organizations, including branches of the U.S. military, picked up and developed Rathbone's insights, creating charts and other tools for decision making.[15]

In the 1960s, companies built electronic devices that could detect looming failures, giving birth to an entire field known as monitoring. But the most significant breakthroughs happened when engineers began to use digital computers for monitoring, data analysis, and prediction. The first of these systems was devised in the late 1970s at Alumax, which was the fourth-largest producer of aluminum in North America. When they built a new smelting plant at Mount Holly, South Carolina, Alumax managers designed a proactive maintenance system. To accomplish this, they had to develop a computer system of their own, since there was nothing like it available in the fledgling database market. Their computer system was innovative: It incorporated a holistic approach to maintenance, as well as the capability to make all plant business functions available to all employees through an online database. It was, in effect, the first computerized maintenance management system (CMMS).

The system's champion at Alumax was John Day, Jr., a man whose name is now revered in professional maintenance and reliability circles. Day was a pioneer in insisting that computers could be useful for maintenance management (this was in the 1970s); for framing maintenance not as a cost but rather as an investment toward profitability; and for demonstrating how maintenance and reliability produced a positive return on investment. Over the years, Day developed his own "maintenance philosophy," complete with elaborate discussions of management approaches, costs, and data for capital expenditures, planned and emergency staff time, inventory, and so on. But perhaps his most enduring contribution was the 6:1 rule, which recommends that every corrective maintenance action should be balanced by six preventive maintenance actions. To put it a different way, for every dollar that a company spends on maintenance, at least 84 cents should be *planned*. Companies that follow this "golden rule" will spend only a relatively small amount on emergency—or reactive—maintenance.

During the 1980s, an era that was obsessed with quality and reliability in manufacturing, the Alumax system received international acclaim from consultants and industry magazines, such as *Plant Engineering* and *Maintenance Technology*. Even today, when professionals talk about "World Class maintenance," they are not referring to any industry standard or published metric. Rather, the term is widely accepted to mean that the vast majority of expenditures on maintenance are planned in advance.

The Alumax system achieved iconic status because John Day and his colleagues were the first to incorporate computer databases and software into their maintenance routines. Subsequent computerized maintenance management systems added sophisticated features for budgeting, cost estimates, inventory and purchasing controls, equipment histories, and data about energy use and conservation.[16] Some systems could check human resources databases to make sure employees were up-to-date on trainings and certifications before assigning them to a task. However, the core challenge for maintenance professionals—one that persists even to this day—was to integrate these systems into everyday operations.

Computerizing maintenance may have made it easier to manage, but, ironically, the widespread diffusion of computers created a whole new layer of technology that also needed maintaining. There's an illusory gap between the promise and reality of digital systems. Computing is surrounded by a set of words that suggest disembodied immateriality, like "virtual" and "cyberspace," but every act we do with digital technologies, from opening an app to searching the Internet, involves some device doing something quite physical, whether in our hand or in a distant "cloud" server.

Again, we're confronted with the paradox: Maintenance is simultaneously needed and neglected. It surrounds us. It's in the back offices of our organizations, stereotypically full of shy, geeky males depicted in shows like the BBC's *The IT Crowd*. It's in repair shops where we take our laptops and phones when they inexplicably crap out. It's in the countless software updates, security patches, and bug fixes that download to our devices. All things built—even digital ones—need maintenance. Do we really understand its importance? Like the Romans who struggled with the sewage that flowed through their streets, we need to make a choice: Are we willing to devote time, energy, and resources to maintenance?

Despite the conveniences and insight that tools like CMMS provide, there is no technological fix that can overcome the absence of a maintenance mindset. Software alone won't save us. We heard a funny illustration of this point when we attended Mainstream, a conference for maintenance managers. A manager who worked at a chemical plant in Arkansas described how, when he got there only a few years earlier, the plant's maintenance was run on the "Billy Bob method." He then pretended to speak into a walkie-talkie, "Uh, Billy Bob, we got a problem here." In other words, maintenance at the plant was completely reactive—it was a response to problems that popped up, rather than anything more organized and planned. By chance, the manager discovered CMMS software that the company had bought *but had never used* sitting on a shelf untouched, like an expensive paperweight.

We see the history of maintenance as a story of barely keeping up, despite some improvements in thinking and technologies over the years. Today, parts of the U.S. rail system are so badly maintained that trains must creep at little more than five miles per hour. Amtrak states that it needs $38 billion to deal with deferred maintenance on the Northeast Corridor rail line between Boston and Washington, D.C., where riders often experience long delays because of track conditions. And in every sector of society, we see how a lack of investment in maintenance is causing catastrophic problems, from dirty hospitals and crumbling bridges to failing schools and inept government agencies.

But politicians, pundits, and executives continue to cry out for more innovation to save us from any number of crises—climate change, economic slowdown, inefficient healthcare, to name just a few. This instinct—to pin all of our hopes on innovation—is exactly the problem that we summarize as the Innovation Delusion.

We will explore the steep costs of the Innovation Delusion in part 2 of this book—the blind pursuit of innovation at all costs, including maintenance of the things that matter most. We'll document those costs at three different scales: societies, in the form of infrastructural neglect; organizations, whose bottom lines suffer from unwise investments in half-baked innovation efforts; and our personal lives, where the constant pressure to "disrupt" is taking a toll on our careers and our time at home.

We've already seen how maintenance is neglected. Now it's time to consider what that negligence is doing to us. The picture isn't pretty.

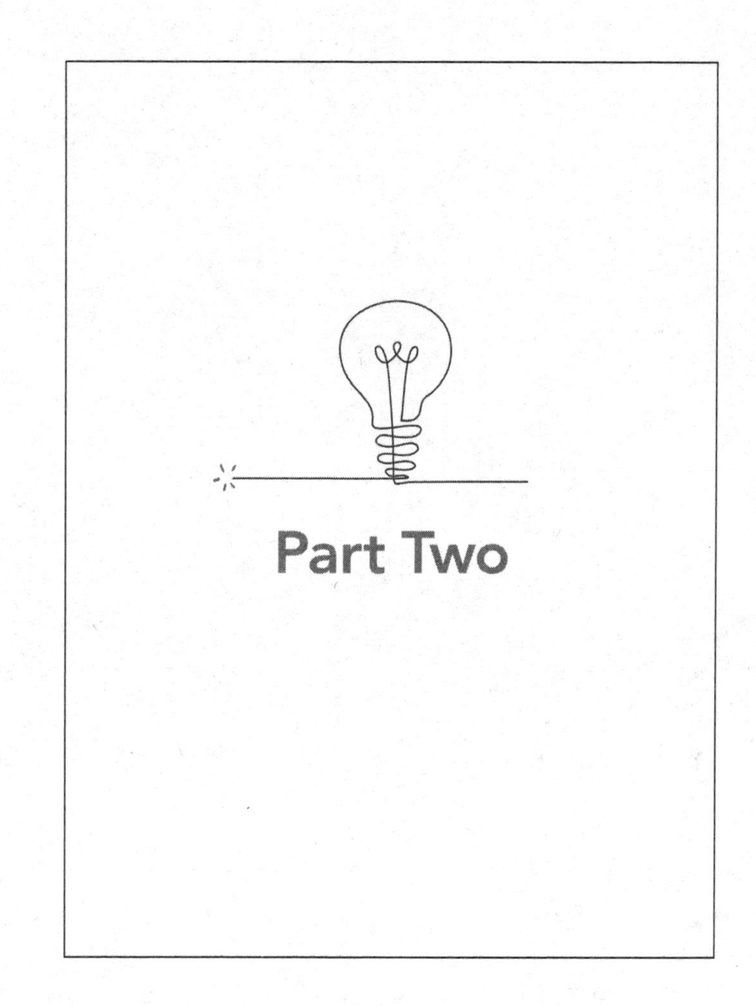

Part Two

CHAPTER FOUR

Slow Disaster

WHAT THE NEGLECT OF MAINTENANCE IS DOING TO OUR INFRASTRUCTURE

In January 2015, riders in Washington, D.C.'s Metro system boarded a Yellow Line train, number 302, and departed L'Enfant Plaza station, heading toward Virginia. Moments later, the train ground to a halt less than four hundred feet from the platform, and the cars filled with thick black smoke. The system's third rail, which carries high-voltage electricity to power the trains, was malfunctioning in a manner known as "arcing." Arcing happens when the insulation in power cables deteriorates, allowing dirt, leaves, trash, and other debris to get through. The electricity grounds there, throwing off sparks and smoke.

On that day, the emergency response was terrifyingly inadequate—almost laughable if the results hadn't been so tragic. It took about forty minutes to cut power to the third rail, and the Metro failed to evacuate train 302 for nearly as long. Some riders fled on their own volition, but many others, especially the elderly and disabled, were stuck in the thickening smoke and darkness. Three passengers gave CPR for twenty minutes to a woman who had collapsed, but she did not revive. A man came, picked the woman up, and carried her away into the smoky blackness, and the three did not see her again. When the smoke cleared, Carol Inman Glover, a sixty-one-year-old mother of two who had recently won an employee of the year award at her job, was dead. More than seventy other people were rushed to area hospitals suffering from smoke inhalation.

In a final report, issued a year later, the National Transportation Safety Board (NTSB) found that Metro Authority had failed "to properly install and maintain third-rail power cables, causing them to become damaged by water and other contaminants."[1] Deferred maintenance—work put off until some later fantasy date when there would be more resources—had killed. What's more, the Metro had failed to adopt safety practices recommended by the NTSB after a 2009 crash that killed nine people—simple things like inspecting tracks, ventilating tunnels and railcars, and sending maintenance workers and fire crews to respond to smoke reports.[2] These safety problems had been well documented ever since the Metro's first fatal accident in 1982. But the Metro appeared to be an organization that did not learn. In its investigation of the 2015 fire, the NTSB found, for instance, that if the Metro had "followed its standard operating procedures, stopping all trains at the first report of smoke, the accident train would not have ended up trapped in the smoke-filled tunnel."[3] No one would have died.

In March 2016, a year after smoke killed Carol Glover, another fire broke out in the early morning hours for the exact same reason—arcing. The problems still hadn't been addressed. The Metro's general manager shut the entire system down for a day to enable emergency inspections and repairs, stranding many riders in the suburbs. The ultimate solution, leaders believed, was to close the worst-maintained train lines down for *weeks*. Only then could the Metro catch up on work that should have been done much earlier.

For years, officials who led and funded the Metro had prioritized technological innovation and system growth over maintaining and caring for what had already been built. From the beginning, planners emphasized automation in designing the system: Trains were computer-driven until a deadly accident in 2009. After that, the technology was discontinued and human operators were put in control.[4] (The D.C. Metro plans to return to computer-driven trains but the process may take five years or more.)[5] Some critics believed that automation and blind faith in technology led to the Metro's lax and lackluster safety culture and ultimately caused accidents.[6] Politicians responsible for the Metro also pushed for constant expansion of the system and construction of new lines, while refusing to levy taxes, create new fees, or raise fares to pay for maintenance, repairs, and other basic operational necessities.

Some of these problems arose from the Metro's unique (and broken) system of governance, which takes orders from an often-dysfunctional board of directors as well as legislators in the city of Washington, D.C., the state legislatures of Maryland and Virginia, and the U.S. Congress. The Metro's financial health, therefore, depends on the goodwill of representatives from, say, Montana and the rural west end of Virginia, who have absolutely no political incentive to ensure that the railway remains in good condition. In fact, showing constituents that they stand against taxation and for limited spending can be a powerful disincentive to allocating funds for maintenance.

We will address issues of governance in a later chapter, but more striking to our eyes is how the Metro's problems were compounded by a mindset that favors superficial ideas of innovation and growth. And it's not the only example of American infrastructure in this position.

In New York City, the summer of 2017 was deemed the "Summer of Hell" after emergency repairs to the subway system snarled traffic and led to long, painful delays. Estimates for what it would cost to fix the system varied between $19 billion and $43 billion over the next fifteen years. A subsequent investigation by *The New York Times* uncovered decades of deferred maintenance and slashed budgets. As was the case with the D.C. Metro, the ultimate cause was a mixed legacy of terrible management decisions and the bad politics of unaccountability. For instance, in 2017, New York governor Andrew Cuomo "pressured the authority to spend tens of millions of dollars to study outfitting M.T.A. bridges with lights capable of choreographed display, install wireless internet and phone-charging ports on buses and paint the state logo on new subway cars."[7] Meanwhile, the MTA cut back on more than forty types of maintenance and delayed routine work on subway cars from every sixty-six days to every seventy-three.

And these are just two examples drawn from transportation. The Oroville Dam in California's Sacramento Valley threatened to collapse in 2017 and send a thirty-foot wall of water down the Feather River, potentially drowning local communities. Two years later, the federal government refused to pay California $306 million for the repairs because it found that the near-disaster stemmed from inadequate maintenance, something federal disaster funds aren't meant to cover. Ten years earlier, in 2007, a bridge holding a section of Interstate 35W collapsed into the Mississippi River in Minneapolis, killing 13 people, and injuring 145. Engineers had already classified the bridge as structurally deficient, but the modest repairs under way at the time of the collapse were, obviously, insufficient to preserve the structure and the lives of the people who depended on it.

If these examples are any indication, it's no wonder that the American Society of Civil Engineers (ASCE) regularly gives the United States near failing grades in its Infrastructure Report Card. Nearly 10 percent of the nation's 613,000 bridges are structurally deficient—meaning that some elements of the bridge require monitoring and/or repair—but the ASCE finds that the country's dams, levees, and drinking water are even worse off, with mass transit sitting in the sorriest shape of all.[8]

The historian Scott Knowles has created a helpful term for describing these situations: slow disaster. Fast disasters, or what we normally just call disasters, include hurricanes, flooding, tornadoes, earthquakes, industrial accidents—events that sweep in quickly, damaging people's lives and the technological systems that support our everyday existence. Fast disasters leave lasting wounds. Long after the story-seeking news cameras have packed up and gone, victims are left picking up the pieces. Some businesses and homes never come back. Lives are shattered.

A slow disaster, by contrast, is the accretion of harm from incremental neglect. It happens when children ingest chips from lead paint or when a potholed road becomes unsafe for traffic.

Slow disasters can lead to fast disasters, of course, when a structurally deficient bridge collapses or poorly tended roadbeds cause a train to derail. And deferred maintenance exacerbates other kinds of disasters, as when a hurricane or an earthquake strikes frail, degraded infrastructure. Hurricane Maria's landfall in Puerto Rico is a terrifying recent example. The U.S. territory had faced years of financial crisis, and maintenance of its electrical system was put off during this period. When the hurricane struck in 2017, it took eleven months to fully restore power on the island. Researchers blamed the failed electricity system for the spike in fatalities that followed the storm, with the final death toll estimated to be three thousand people.

The notion of slow disasters brings the issue of deferred maintenance and the way we prioritize new systems into focus, in part because it helps us to see these stories in the long view. It snaps us out of the immediacy of the twenty-four-hour news cycle—with foolhardy journalists dancing in front of cameras to show us how hard hurricane winds blow—and draws our attention to the long-term damage of malignant neglect.

We must be careful when talking about infrastructure problems. The United States—and other rich nations—have experienced declining or stagnant growth, productivity, and wage gains since the 1970s. Too often people explain these changes through overly simple anecdotes about moral decline. For example, productivity decreases in the 1970s were sometimes blamed on lazy, long-haired hippies smoking too many j-birds and lacking a work ethic. Blaming bad events on moral failure is easy pickings, but often it's not the best explanation when things go wrong.

Some thinkers, such as the conservative commentator David Stockman, who led the Office of Management and Budget under Ronald Reagan, believe that the widespread worries about infrastructure are considerably overblown. We disagree with his assessment, but we share his belief that we should always be on guard for hyperbole and hysteria. Serious infrastructural troubles are demonstrable, and there is zero reason to believe they'll improve anytime soon.

Moral turpitude, stupidity, and lack of wisdom may sometimes—maybe even often—explain infrastructural decay, but the causes are extremely complex and as varied as problems that can arise. At the core, however, we believe a (sometimes literally) deadly gumbo of factors has led to our current situation, including the way we value innovation and growth over caring for the world we have already built.

FRAGILE FOUNDATIONS

It is difficult to estimate the enormous contributions that modern infrastructure has made to improving our everyday lives. Water and sewer systems provide clean water, remove waste, and keep us safe from miserable diseases like cholera, hookworm, and dysentery, which killed humans for most of history and still do in many places on this planet. Electrical systems power a vast range of technologies that make our lives easier and more comfortable. Communications systems—from telephones to the Internet—connect us to other people in ways that were simply unimaginable a few hundred years ago. So many aspects of what we think of as modern life ride on the backs of the technological networks we call infrastructure.

As we saw in chapter 3, people in our culture have become fixated on innovation-speak—in part, because of how new technologies have contributed to economic growth. They enable us to do more with less, and transportation technologies, along with the infrastructure that supports them, provide a clear example of how this works. When the United States was founded, horses were the predominant mode of transportation. They traveled down dirt roads that were often poorly cared for, and muddy to the point of being impassable during some seasons. It took days to get from New York to Virginia. Travel was so bad that, when James Madison advocated for the formation of the federal government in the *Federalist Papers,* he argued that the country did not have to worry about the formation of violent factions or dangerous demagogues. Communications were simply too slow for people to whip one another up into a frenzy across hundreds of miles.

Clearly, transportation has improved since then: railroads; automobiles, semitrucks, modern concrete roads, and eventually the Interstate Highway System; aviation and intermodal transportation hubs; and since the 1970s, the rise of enormous container ships and the global production networks they enable. Even by the mid-twentieth century, it was possible for a screwdriver to be manufactured in the Midwest one day and used in New York City the next. The improvement in productivity since 1800 has been staggering.

But what degrading infrastructure makes clear is that economic growth is not unidirectional. If human activities can improve and become more efficient, they can also degrade and become less productive. When it comes to infrastructure, this truth can perhaps be seen most clearly by thinking about beer.

One of our friends worked for years for the famous American brewery Sierra Nevada. Because the American railroad system is so badly maintained and train cars rock back and forth so violently, Sierra Nevada and other breweries have resorted to adding extra padding and cushioning to their shipments to protect their bottles from breaking. This padding material takes up space, so the companies ship less beer than they could under better circumstances. Moreover, hundreds or thousands of miles of American railroad track is degraded to the point that trains must travel slower over some stretches than they should be able to. All of this means that railroad transportation is *less* efficient than it could be—and, in fact, once was. We have lost out on growth and put the well-being of our alcoholic beverages in jeopardy. Which raises a question: Have we so lost sight of our basic human values that we are even willing to risk our *beer*?

Worries about deteriorating infrastructure are as old as the large-scale technological systems that constitute them, whether privately owned, like railroads, or public, like waterworks. But economists, policy analysts, and civil engineers have published a steady stream of reports highlighting infrastructure troubles ever since the 1970s. That timing is not surprising. When the U.S. economy hit the skids in the seventies, governments at the local, state, and federal levels tried to cut costs. Maintenance was a near-constant victim.

Pat Choate and Susan Walter shined some early light on this mounting problem in 1981, when they published a report for the Council of State Planning Agencies titled *America in Ruins* that covered all kinds of public infrastructure from sewer systems to highways to subways. For example, it found that the cost to rehabilitate nonurban highways would exceed $700 billion during the 1980s, and if inflation of maintenance costs continued at the prevailing rate of 12.5 percent, current funding levels would only cover about one-third of the bill.[9] Another finding was that the public works in New York City alone would require more than $40 billion in maintenance over the next decade to bring systems up to a state of good repair. As we now know, as far as the subway was concerned at least, the money was never allocated.

As Choate and Walter argued, post-slump budget cuts undermined "efforts to revitalize the economy and threaten[ed], in hundreds of communities, the continuation of such basic services as fire protection, public transportation, and water supplies."[10] While the report originally got little coverage, it was eventually highlighted in *The New York Times, Time,* and *Newsweek*. Today, Choate and Walter's warnings sound prophetic.

Three years after *America in Ruins* came out, Congress created the National Council on Public Works Improvement, which was led by influential figures drawn from business and government and given a mandate to produce a report on the state of American infrastructure. Published in 1988, *Fragile Foundations: A Report on America's Public Works* was on balance probably more focused on creating new infrastructure than caring for existing systems. But maintenance continued to be an important theme, in part because spending on public works was falling at the same time that maintenance costs were expanding at a rate that easily outpaced inflation. Operations and maintenance costs had risen from $21.6 billion in 1960 to $56.5 billion in 1984, while public spending on infrastructure had dropped from 3.6 percent of gross domestic product in 1960 to 2.6 percent in 1985.[11] As spending on maintenance continues to decline, more existing bridges, dams, levees, and other structures will crumble even as we continue to build new ones.

Fragile Foundations was one of the first texts on infrastructure to name a dynamic that would later become a constant theme: Maintenance isn't sexy. As the report noted, "Maintenance spending does not generate the excitement associated with new capital projects. The public is seldom aware of maintenance unless a pothole persists or a bus air conditioner breaks down. Along with being invisible, maintenance is not politically compelling. . . . Operation and maintenance budgets are often easy to cut because voters do not see infrastructure deterioration."[12]

The most lasting influence of *Fragile Foundations,* however, is that it was the first infrastructure publication to include a report card. The report graded eight types of public works, with water resources receiving the highest grade, a B, and hazardous waste getting the lowest, a D. Mass transit got a C- in part because "maintenance has been erratic and inadequate, especially in older cities."[13]

A decade later, leaders at the American Society of Civil Engineers were looking for a project that would boost the organization's profile. When it realized that Congress was not going to update *Fragile Foundations,* staff members hit upon the idea of putting out a report card of their own. First issued in 1998 and titled Report Card on America's Infrastructure, the report card has been issued about every four years since. If the ASCE's goal was to get attention, the report card has been a massive success. Bill Clinton referenced the 1998 report card's F grade for public schools only a few days after it was published, and Barack Obama used the 2009 and 2013 report cards when arguing for increasing infrastructure spending.[14] Most major news outlets in the United States have cited the report card. It has been hugely influential in generating public conversations about the nation's infrastructure needs. As the engineering society put it, "In each of ASCE's six Report Cards, the Society found that these same problems persist. Our nation's infrastructure is aging, underperforming, and in need of sustained care and action."[15]

The answer, infrastructure advocates argue, is increased spending. We agree. But there's another way of thinking about infrastructure that has come to light in the last decade. If it is right, our situation is far more dire than even the infrastructure boosters believe, and no amount of spending is going to get us out of the hole we're in.

POSTPONING REALITY

Charles "Chuck" Marohn is a straitlaced, even square, civil engineer. A soft-spoken Catholic and registered Republican, he grew up as a farm boy and served in the National Guard before marrying his high school sweetheart and moving to a small town in the largely rural Midwest. All of this makes Marohn an unlikely candidate for the title "thought leader." Yet, from his hometown of Brainerd, Minnesota, Marohn and his colleagues have started a growing, influential movement called Strong Towns, a nonprofit that works to make American cities financially resilient.

After graduating from college, Marohn was a typical civil engineer, working to develop communities around his region. "I built sprawl" was how he later told it to an interviewer. Beginning in his mid-twenties, however, Marohn began to experience a crisis of faith. He felt frustrated at work. His ambitions to move up in the organization significantly outpaced the promotions his bosses were willing to give him. He'd likely be working for decades at his current level before he'd be able to climb the ladder.

Then he got an opportunity to take part in a foreign exchange program with the local Rotary club, which sent him and some others to Italy. When Marohn and his traveling companions arrived, they found that the Italians were not prepared for them. The trip fell apart, and the Rotary club recalled Marohn's group. But Marohn refused to leave. He rented a car and drove around Italy for more than a month, sleeping in the vehicle at night.

During his travels, he paid special attention to the country's infrastructure and watched how construction crews worked. He was interested in what he saw, but he also brought with him a sense of American superiority. He couldn't believe how primitive some of the Italian engineering practices were. "There was this one place," he told us, in Lecce, a town in the boot heel of Italy, "where they were fixing a pipe, and men were bringing these big rocks out of the ground—the stones that were the street. My first reaction was the reaction of a twenty-five-year-old American, which is, 'These people are dumb.' Like, 'Look at these fools sitting here manually lifting rocks with their hands. What a bunch of idiots.' But then once I humbled myself . . . I started to realize, like, OK, in the United States we build a paved road, and it lasts, like, twelve years or it falls apart unless you do intense maintenance and then it will last maybe twenty-five or thirty years. That friggin' rock has been there since A.D. 400."

Marohn had a hard time functioning when he got home from that trip. Something had changed in him. At the age of twenty-six, he considered quitting his job, getting a divorce, and leaving his hometown. While he struggled with this existential crisis, he experienced something of a professional epiphany.

Marohn had been put in charge of a project in Remer, Minnesota, a town of less than four hundred people that had been fined by the state for discharging too much wastewater. Remer's wastewater retention and treatment ponds were overflowing, and, as Marohn later wrote, the spillover was "threatening to collapse the earthen berm and dump thousands of gallons of concentrated sewage into the adjacent [Willow River]."[16] In cases like this, the usual problem is that sewer pipes are letting in fresh groundwater, which causes the system to overflow. Marohn investigated, "going from manhole to manhole during the middle of the night," testing flows to see if he could identify the leak. He found the culprit. It was a three-hundred-foot pipe that ran directly under a nearby highway. Repairs on the pipe would cost $300,000, but the town's entire annual budget was only half that. Remer did not have the money to maintain its own systems.

Marohn looked around for government aid, but no federal grant program addressed such small projects—*especially* not projects focused on maintenance. So Marohn dreamed up a clever solution. He designed an upgraded and greatly expanded water system, in which the original repairs were included almost as an afterthought. The new system would cost $2.6 million. The project "was now perfect for grant programs," Marohn later wrote.

Marohn put together a successful grant application and got the system funded, though it required Remer to get a $130,000 Department of Agriculture loan that the town had no business taking on.[17] The townspeople were ecstatic. Politicians lined up for ribbon-cutting photo ops and even declared a "Chuck Marohn Day." "It wasn't like a parade or anything," Marohn told us. "It was a little tent with a grill and some hot dogs, and people sat around going, like, 'This is awesome. We fixed this city. This is great.'" He also got a "nice bonus" out of the experience. But as time passed and Marohn thought about what he'd done, he came to believe that there was a fundamental lie built into the heart of American infrastructure policy.

Here's the basic problem: Federal funding supports a considerable amount of new infrastructure built in the United States. But it is difficult, sometimes impossible, to use federal money to *maintain* these systems. For example, in 2014, the federal government paid for nearly 40 percent (or $69 billion) of new infrastructure projects but only 12 percent (or $27 billion) of operations and maintenance.[18] Put another way, more than 70 percent of federal infrastructure spending went to new construction, whereas 65 percent of state and local funding went into operations and maintenance.[19] Localities are happy to accept funding for new development, even though they are implicitly agreeing to perform maintenance for the life of the system that's being built.

None of this would be a problem if American communities were generating enough tax revenue to cover maintenance costs—but as a rule, they aren't. Not even close. For instance, when Marohn and some colleagues did a study of Lafayette, Louisiana, they found that the city had infrastructure needs of about $32 billion but a tax base of only $16 billion.[20] The average family in Lafayette paid $1,500 a year in taxes, of which about 10 percent went to infrastructure maintenance. Marohn estimated that each family would need to pay $3,300 in additional taxes each year "just to tread water," without even adding new roads or other structures or significantly improving existing ones.[21] Most families simply couldn't afford this, and the community's leaders would not be in office for long if they tried to raise taxes in line with reality. Marohn believes that Lafayette's upside-down financial situation is becoming the norm in communities across the United States.

Projects like the one in Remer have burdened localities with extravagant infrastructure that they can't afford. The consequences lie over the horizon, however, so officials and citizens can congratulate themselves for accomplishing something while leaving the worries and problems to the future. You can see how radically different this way of doing things is from what Marohn witnessed in Italy, where infrastructural labor was slower—even less efficient—but demonstrably sustainable. After all, it had been going on for millennia.

The tension between these two ways of doing things became too much for Marohn. Thinking about leaving his career, town, and marriage, he fantasized about working at something "simple and happy." "I wanted to be a gondola driver or drive a bus at Disney World," he told us. As Marohn planned for the future, his wife asked him to figure out a solution that didn't involve divorce.

Marohn ultimately decided to get a master's degree in urban and regional planning at the University of Minnesota. He learned a lot in the program, and much of it conflicted with what he had learned earlier as an engineer. For example, in his old job, Marohn had helped lay out subdivisions with curving lanes and cul-de-sacs. But in planning school, he learned that the traditional grid organization of streets was better for communities because it enabled much more efficient travel and flexible development patterns. When he graduated and returned to Brainerd, he opened up a consultancy called the Community Growth Institute, which helped small towns make decisions about planning, codes, and zoning. "Our internal mission was to save rural America," he told us.

There was still a tension between what Marohn had witnessed in places like Remer and what he had learned in graduate school, but he didn't see it yet. "I think that . . . every planner believes that if you just had the right set of zoning regulations that you can solve every problem. Like, you can cure cancer and have world peace. . . . It's seductive. You start to believe that you have way more knowledge and way more insight than other people."

Marohn was open-minded enough to realize he could be wrong, and he had an intellectual awakening after reading Malcolm Gladwell's essay "Blowing Up." Gladwell contrasts two different investors, Victor Niederhoffer and Nassim Nicholas Taleb, the latter of whom went on to write bestselling books like *The Black Swan* and *Antifragile*. Niederhoffer was in many ways a traditional investor who believed that you could find opportunities for profits in a market through mathematical analysis. In the 1980s and early 1990s, he raked in cash using this method.

Many people attributed Niederhoffer's success to his expertise, presuming he possessed some form of knowledge that others couldn't access. But Taleb took a completely different approach. He assumed that he was fundamentally ignorant and could *not* predict the future—that tomorrow was and is much more uncertain than he could possibly estimate. He used options to bet on dramatic swings in the market, wagering that things would change in ways no one could anticipate.

In Gladwell's telling, the moral was clear. Niederhoffer's investment company tanked and was dissolved in 1997 after taking heavy losses—and his next company failed about a decade later. Taleb's method of building robust, or resilient, strategies was superior, because it would not be undermined by unexpected negative events.

Marohn saw a deep truth in Gladwell's essay and believed it raised fundamental questions about his own professional fields of planning and engineering. Planners, he reasoned, had been trained to think like and see themselves as Niederhoffers—individuals who should foresee and control the future. As a result, they radically underestimated the complexity of human communities. It was better, Marohn came to believe, to see things like Taleb: avoid overplanning and choose simple but resilient solutions. The status quo of loading up localities with infrastructure they could in no way afford to maintain was the definition of fragility, the dark antipode of resilience.

Marohn came to believe that this problematic situation arises partly from human psychology. He has a diagram of cognitive biases hanging on the wall of his Brainerd office. Americans, he believes, have a tendency to discount future costs in light of present gains. Perhaps other psychological factors, including a propensity to favor simple stories and solutions and a habit of assuming that the future will be like the present, cloud our thinking around issues like city and infrastructure planning.

As Marohn's thinking developed under the influence of writers like Taleb and the libertarian economist Friedrich Hayek, it created tension with his role as a planner. "I became self-destructive in a professional sense," Marohn told us. "I would show up to a city, and they'd be like, 'Well, we would really like to build a community center so we can get our youth together to play ping-pong.' And I'm like, 'What the fuck are you talking about? This building is crumbling, you can't fix it! How stupid are you people?'" But something intervened to spare him the pain of fleeing the planning industry as he had earlier run away from civil engineering—failure.

Business had started to slow down for the Community Growth Institute in 2006 and 2007 at the dawn of the financial crisis that took hold in 2008 and 2009; the building and construction work that the institute depended on was ravaged. Marohn had to start laying people off, and by the time the crisis hit the rest of the country, the company was effectively dead—the only thing left was the debt Marohn owed.

Before the institute folded, however, Marohn had started writing about his worldview in something he titled "The Planner Blog." He did it not only because he thought it would be therapeutic but also because it would enable him to get his thoughts down in a way that could be useful for others. "We were going to cities [that were] going bankrupt," he told us. He wanted a concise way to encourage municipalities to be resilient and financially sound—a state of being that he started calling "strong towns."

The blog focused on local issues at first, but it started reaching more readers as people passed around Marohn's contrarian takes. Eventually, a friend encouraged Marohn to start up a nonprofit to communicate his ideas. Marohn told his friend he didn't know what he would call such an organization. "And I remember him being like, 'Are you stupid? We call it Strong Towns, of course.'"

Since that time, Strong Towns has become a major voice in rethinking the problems of American infrastructure and the deferred maintenance that cripples it. Marohn looks to history to explain how we got in this position. For hundreds of years, humans built towns and cities in traditional, relatively dense, walkable forms. Such places were not only human in scale but also economically productive: They generated enough tax revenue to cover the costs of their systems. The streetcar suburbs that developed beginning in the late nineteenth century were less dense than cities but still fairly compact compared to what came later.

The real change came after World War II with what Marohn calls "the great American experiment of suburbanization." Postwar planned subdivisions, like the famous Levittown on Long Island, presumed that their residents owned automobiles. They were, in the scope of human history, extremely low-density "towns," and yet they also required more intense infrastructure: more roads, more sewer and water pipes, more utilities.

Intense political and economic power pushed for the spread and continuation of this model. The combined interests of developers, construction companies, real estate brokers, and related industries formed a powerful lobbying force at the state level—and still do to this day. Few laws impede the desires and plans of this crew. Often aided by money from the federal agencies as well as tax breaks from nearly every level of government, developers and contractors built subdivisions with intensive infrastructure and then handed over the maintenance to municipalities.

In post after post on the Strong Towns blog, Marohn has examined how this unsustainable situation developed. In his view, the initial postwar suburban boom pales in comparison to what happened next. Beginning in the 1960s, white flight and other factors hollowed out the economically productive city centers throughout the country, leaving dead downtowns that were largely devoid of life outside business hours. Infrastructure has rotted in the so-called Rust Belt and in cities around the country where tax bases have evaporated.

Yet, even in the face of a stagnating economy and diminishing tax bases, politicians in towns and cities chased growth, both by using federal funding and by borrowing money. At the end of World War II, municipal bond debt added up to 1 percent of gross domestic product (GDP). That number had shot up to 6 percent by 1980. Today it stands at 27 percent of GDP.[22] Politicians have every incentive to take on debt—whether in the form of bonds or the form of infrastructure that will need to be maintained down the road—in the name of economic growth. Growth means jobs and money and pretty new things. Politicians need to look like they are doing something, and they don't pay the price for infrastructure whose downstream maintenance costs only become clear after a decade or more.

Marohn also criticizes urban planners—and engineers, especially—for encouraging a pro-growth mindset that focuses almost exclusively on building new infrastructure. He calls the American Society of Civil Engineers' Infrastructure Report Card and related efforts the "infrastructure cult" because, in his view, they are based on an irrational, unfounded faith in growth and the idea that more is better. Marohn joked that one ASCE report should have been titled "Pretending it is 1952." His criticisms have often made engineers unhappy. In early 2015, Marohn received notice that a former ASCE fellow and Minnesota civil engineer filed a complaint with the state against Marohn's engineering license, alleging "misconduct on the website/blog Strong Towns."[23] The complaint came after Marohn pointed out that civil engineers in Minnesota had an obvious conflict of interest when pushing for expanded infrastructure spending.

You don't have to share Marohn's conservative penchant for limited government to appreciate his basic insight: If governments, organizations, and individuals build and buy systems without providing for their future care, we end up facing a stress-inducing mountain of deferred maintenance and infrastructural debt, which is precisely what we see in many parts of society today. Over the past six years of research and conversations with experts in multiple fields, we've come to believe that Marohn's framework can be applied to other problems of technology and maintenance. From libraries to corporations to single-family homes, individuals and groups take on technologies—sometimes with great glee and excitement—without thinking about the long-term obligations they entail.

Marohn's work also helps us make a key distinction when discussing worries about infrastructure: President Donald Trump promised to improve American infrastructure during his 2016 presidential campaign, and infrastructure has been a frequent topic in the years since. (Progress on any meaningful legislation remains elusive.) But most of this chatter, including Trump's original plans, focuses on building *new infrastructure,* not maintaining and fixing what we already have. Moreover, "maintenance" often functions as a euphemism for widening streets, adding more sophisticated traffic-control technologies, and making other upgrades. Some of these changes no doubt benefit the public good, but they also increase the infrastructural debt by creating even more stuff that needs to be maintained and repaired. Put another way, even when we hear a great deal about infrastructure in public discourse, true maintenance is rarely the focus. And this neglect levies disproportionate costs on populations that have already borne the brunt of systemic patterns of social and economic inequality.

HOW THE OTHER HALF LIVES

In the wake of the Flint water crisis, which began in 2014, journalists found that thousands upon thousands of American cities suffered from elevated levels of lead. Reuters uncovered nearly three thousand areas with lead levels higher than Flint's—with a combined population of roughly 12.5 million people. In more than a thousand of these communities, blood lead levels were four times worse than those in Flint at its nadir. As the journalists M. B. Pell and Joshua Schneyer put it, "Like Flint, many of these localities are plagued by legacy lead: crumbling paint, plumbing, or industrial waste left behind."[24] People are living in the toxic shadow of mistakes that should have been undone long ago but for want of money, resources, and care.

A major problem facing Flint and many other lead-poisoned towns is depopulation. Chuck Marohn highlights how American infrastructure policy often assumes growth—our infrastructural choices are marked by a kind of naïve, even dangerous, optimism that the people of tomorrow will somehow be able to pay for structures we build today. But often enough presumptions of growth get things exactly wrong. The plain reality is many American municipalities have been experiencing shrinking populations for decades, making their infrastructural struggles all the more punishing. Depopulation undercuts cities' tax bases, leaving them with fewer and fewer resources to deal with problems, including keeping up with infrastructure maintenance. These communities face wrenching—sometimes perilous—choices.

Lead-contaminated water is not even close to the only problem that depopulating cities face. The city of Baltimore hired Rudy Chow in 2011 to manage its Bureau of Water and Wastewater, which includes not just the city of Baltimore but also surrounding counties. Born in Taiwan, Chow came to the United States as a teenager and ended up studying engineering in college. For twenty-seven years, he worked on water issues at the Washington Suburban Sanitary Commission, an organization that manages more than ten thousand miles of fresh water and sewer pipelines in Maryland's Prince George's and

Montgomery counties. After he retired, he joined Baltimore's Department of Public Works. Following what many people considered to be a successful run at the Bureau of Water, he became the director of the Department of Public Works.

When Chow first came to Baltimore, the city's water system was in bad shape. He was impressed by the network's overall design, but the water pipes themselves had suffered from decades of deferred maintenance. The system had gone through an extended period of expansion and growth from the 1960s until perhaps as late as the 1990s. "But there was no long-term plan for how to care for the system," Chow told us. And, in this, Baltimore isn't alone. "I talk to water managers all over the country—and this is the problem they all face."

A simple way to measure the problems of water systems in older cities like Baltimore is to count the number of pipe failures—in the case of the Charm City, about twelve hundred a year. When a pipe fails, streets and sometimes residences flood, and public works crews have to dig up the pavement to repair the break. Sometimes the problems in Baltimore get even worse. In February 2018, during a particularly brutal cold spell, the city experienced six hundred pipe failures—nearly half of its annual average—in a single month. The Department of Public Works was forced to put its crews on mandatory emergency shifts: sixteen hours a day, six days a week, for weeks on end. Baltimore's infrastructure problems were completely unsustainable.

In chapter 9, we will learn about the plan that Rudy Chow and other leaders at Baltimore's Department of Public Works have implemented, and why they believe the future will be better, perhaps even hopeful. Still it's important to realize that in some ways Baltimore and cities like it are lucky: Having crumbling, even failing, infrastructure is sometimes better than having none at all.

The simple fact is that many people in the United States have never had access to technological systems that define modern experience. As the science fiction writer William Gibson once put it, "The future is already here—it's just not very evenly distributed." This point was dramatically highlighted in December 2017 when Philip Alston, a special rapporteur on extreme poverty and human rights at the United Nations, toured parts of the United States to examine the conditions under which the nation's poorest people lived. Alston had previously inspected countries like Mauritania, Chile, and Romania, places that everyone knows are poor. But he reported being shocked by what he found in the United States.

In Alabama's Black Belt, a region named for its fecund, dark topsoil, Alston met people who had unreliable electricity service and completely lacked modern sewer systems.[25] He toured one small home where five members of an extended family, including two children and an eighteen-year-old with Down syndrome, lived. For plumbing, that house and the ones around it relied on so-called "straight pipes": PVC rigs that dump raw sewage into open-air pools, filling the air with a terrible odor. But the problem was more than merely aesthetic. Because the family's main water line was also in a state of disrepair and likely had cracks, human waste was getting into their drinking water. "Everyone gets sick all at once," one of the adults told Alston.

Months earlier, a scientific investigation had found that 34 percent of individuals living in the county tested positive for genetic traces of *Necator americanus,* or hookworm. Hookworm is an intestinal parasite mostly associated with poor, tropical nations and thought to be virtually eradicated in the United States. "I think it's very uncommon in the First World," Alston told reporters while touring Alabama.[26] Hookworm infections occur when skin comes into contact with raw sewage. The parasite attaches itself to the host's small intestine and feeds on blood, which causes negative health effects like "iron deficiency and anemia, weight loss, tiredness and impaired mental function," especially in developing children.[27] A survey found that 73 percent of people in Lowndes County had been exposed to raw sewage, either on the ground or when it washed back into their houses from failed septic tanks and degrading water management systems.

Ed Pilkington, a journalist from *The Guardian,* went to see for himself. At one trailer park, he saw a PVC straight pipe run from a mobile home to some trees only thirty feet away. The pipe, which was cracked in several places, ran within a few feet of a basketball hoop where children played. "The open sewer was festooned with mosquitoes, and a long cordon of ants could be seen trailing along the waste pipe from the house," Pilkington wrote. "At the end of the pool nearest the house the treacly fluid was glistening in the dappled sunlight—a closer look revealed that it was actually moving, its human effluence heaving and churning with thousands of worms."[28]

This reality, along with Chuck Marohn's assessments of American infrastructure, raises enormous political and moral questions for our society. For example, should we view clean water, stable bridges, and sewer systems that don't cover streets with human feces a human right? If so, how should we pay to make these systems universally available? How much should large cities and rich communities be asked to pay to prop up poor rural districts? Who is ultimately responsible when existing infrastructure undermines public health?

In the *Freakonomics Radio* episode "In Praise of Maintenance," which was based in part on our work, the host Stephen Dubner asked Lawrence Summers, the famed economist and former U.S. Treasury secretary and president of Harvard University, about the tension between innovation and maintenance. Summers responded, "I think a great nation can walk and chew gum at the same time." We wish that was true. But when we examine the state of infrastructure in the context of our culture laden with innovation-speak, we find that parts of our society are falling behind while others obsess over racing ahead. In the next three chapters, we find something similar in other domains of our society, from how leaders invest in businesses and universities to the lived experiences of workers and the ordinary maintenance of keeping house. Like infrastructure, those domains struggle with maintenance mainly because they obsess over growth and short-term gains instead of taking care of what they have.

CHAPTER FIVE

Growth at All Costs

THE INNOVATION DELUSION IN BUSINESS

When Jeffrey Immelt became CEO of General Electric on September 7, 2001, the corporation regularly topped polls of the world's most admired companies. Immelt's predecessor and mentor, Jack Welch, had become an archetype of business school case studies—the no-nonsense executive who "rightsized" and modernized the manufacturing giant, helping it become a leader in the lucrative markets for financial services. But Immelt's first decade left a lot to be desired. GE's stock lost roughly half its value between 2001 and 2011. And despite incremental growth for the company over the next few years, Immelt was clearly impatient for more. In a June 2015 speech at The Economic Club of Washington, D.C., he worried that "the U.S. is growing too slowly." Fortunately, the solution was close at hand: "Almost all of our problems can be solved with stronger growth."[1]

Since Immelt openly admired the "great companies like Apple, Facebook and Google," which led the global economy into the digital age, it was not surprising to see GE align its public image with theirs by going all in on innovation-speak. A 2016 *New York Times* profile of the company's new software division left readers with an impression of a crowd of grandfathers crashing an episode of *Silicon Valley:* "Employees companywide have been making pilgrimages to San Ramon for technology briefings, but also to soak in the culture. Their marching orders are to try to adapt the digital wizardry and hurry-up habits of Silicon Valley to G.E.'s world of industrial manufacturing."[2] It referred to General Electric as a "124-year-old software start-up."

It was around this time, in the fall of 2016, that we received an unsolicited email from a manager at GE. Our correspondent, whom we'll call "Brian," mentioned that he had read an essay of ours exploring some of the ideas in this book and found that our criticisms hit close to home. He wrote that GE was rallying around the concept of "innovation," and asked us if we could help his team think through some of the "historical contexts and counterpoints to the hype." We were fascinated: Here was a manager in a company that had drunk the innovation-speak Kool-Aid in an earnest attempt to recapture their mojo, and he wanted to talk to us. How could we refuse?

We soon scheduled follow-up calls with Brian, who asked his boss to join the conversation. We learned that GE was embracing a "fail fast" mentality and encouraging internal business units to think and act like entrepreneurs—precisely like the "124-year-old software start-up" that *The New York Times* described. We eventually visited Brian and a few dozen of his colleagues at a team retreat. We were the lunchtime entertainment, so we scarfed down sandwiches and then launched into our slides. In our presentation, we warned Brian's team about several of the problems we've already outlined in this book: that Silicon Valley's "fail fast" mantra has limits as a general model; that "disruption" can cause pain and damage in people's lives without delivering the promised results; and that leaders should not lose sight of the fundamental importance of maintenance, reliability, and the hard work it takes to keep things running smoothly and safely. An engaging period of questions and discussion left us feeling like the audience had understood and appreciated our message. But as we packed up to go, we wondered how the team might act on our overarching recommendation.

On the way out, we couldn't resist asking Brian how he found our contrarian essay, "Hail the Maintainers," that prompted him to reach out to us in the first place. "Oh yeah," he said, then laughed as he launched into a story about a day when he suffered through hours of meetings about the benefits that GE would capture by becoming more innovative. The meetings finally ended, mercifully, but the agony stayed with him as he surfed the Web late that night. In a digital cry of despair, he banged the words "FUCK INNOVATION" into a search engine and—voilà!—our names appeared.

Even if our intervention changed some hearts and minds at GE, the company's strategic trajectory was already firmly fixed. A year later, our concerns proved accurate: GE was not delivered to the promised land by its attempts to become more agile and entrepreneurial. In June 2017, the company announced that Immelt would step down as CEO. By the end of the year, its stock price fell from $27 per share (same as it was when we visited in November 2016) to $16.90. As 2018 rolled on, GE's tailspin continued to attract negative press—one writer tracked GE's journey from "American icon to astonishing mess," and another probed "how General Electric became a general disappointment." At the end of 2018, GE's stock price was $7.17.

If this behavior was restricted to a single firm, then we could shrug our shoulders and walk away. But GE is not alone in believing that success is simply a matter of tricking its employees into acting like they work for a software start-up, or that a manufacturing giant can mimic the approaches of younger, smaller, and nimbler companies. The problem runs much deeper. This way of thinking—that salvation will come if we glom onto the latest trends, become more "innovative," and grow faster—has spread throughout American business and into essential public organizations.

This chapter will explore the impact of the Innovation Delusion on institutions, including businesses, schools, and hospitals, that structure some of the most important aspects of our lives. These stories provide a more vivid picture of the price of neglect, the pressures of long-term decline, and the dangers of buying into fantasies of renewal and endless innovation. The same kinds of problems and the same patterns detailed in chapter 4 are evident: the preponderance of superficial ideas of innovation and growth; the political risks involved with (responsible) investments in maintenance; and the fact that neglecting maintenance often brings disproportionate harm to people already grappling with social and economic disadvantages.

THE GOSPEL OF GROWTH

It is impossible to overstate the importance of *growth* in our collective beliefs about business, education, health, and even the world of ideas. To thrive is to grow, whether raising a child or managing personal finances is at stake. It's more than an aspiration; in many cases, it's an instinct. The gospel of growth is a part of economic orthodoxy, deeply ingrained in all varieties of industrial capitalism.

Growth is the fuel for both a company's financial health and a society's economic well-being—which is why journalists and elected officials invoke measures like gross domestic product or the Dow Jones Industrial Average to show that things are moving in the right (or wrong!) direction. If GDP is up, or there are gains in the Dow, it is a good day—because societal progress can be measured by gains in productivity, economic growth, and the spreading of material abundance among an ever-expanding number of citizens. The logic of progress, when seen through the lens of GDP or the stock market, is very simple: Things get better when we accelerate productivity, economic growth, and material abundance. In other words, for any conceivable problem, growth came to be seen as the *only* solution. As the historian Eli Cook put it, "American society's top priority had become its bottom line," and "net worth was synonymous with self-worth."[3]

Americans readily embraced this logic, but seem to have been slow to fully grasp the implications that the only way to create clean, simple measures like GDP or the Dow is by leaving more complex variables out of the equation. As a result, they ignore values that aren't easily measured—such as misery, inequality, and joy—and narrow the aperture for appreciating the things that make life worth living. It's a strange world where all things that have value—land, labor, technology, ingenuity, affection, joy, misery, and so on—are reduced to quantities that can be plugged into balance sheets and judged by their utility for producing profit.

Growth is a double-edged sword. Physicians use the word "obesity" to describe unconstrained growth that comes from poor dietary choices. And political scientists use the word "empire" to describe unconstrained geopolitical growth. In both of these examples, growth occupies a paradoxical position: Although it is the basis for many positive outcomes, there is hell to pay when growth triggers a never-ending cycle of more growth for growth's sake. These ideas are not new to economists, and concerns about the "fairytales of eternal economic growth," in Greta Thunberg's moving phrase, have been around for centuries. The founders of modern economics, including Adam Smith and John Stuart Mill, recognized that limits to existing supplies of land and materials meant that economic growth would diminish naturally over time. Yet the growth delusion persists because humans instinctively revert to short-term thinking and respond to simple, short-term incentives.

In chapter 4, Chuck Marohn describes what happens when delusions of growth are mindlessly applied in the realm of infrastructure. The same dynamic occurs in the world of business: Entrepreneurs build shiny new objects, but they and their investors rarely plan for the days when those objects lose their luster and the company staggers under the weight of deferred maintenance and technological debt. As we saw with GE, executives and managers who find themselves in this situation are tempted to turn to buzzwords and trends in innovation (Big Data! Automation! Blockchain!) that promise to solve all problems by generating easy, endless growth.

It's important to emphasize that these executives and managers aren't evil, and they aren't dupes. Rather, they are trapped in situations in which their primary duty is to ensure that investors and shareholders are happy—in other words, that the corporation creates ever-increasing returns on investment. Scholars and politicians alike have wrestled with the thorny problem of putting "shareholder value" above all other outcomes. For example, the economists William Lazonick and Mary O'Sullivan have shown how companies that fixate on growth have a way of undermining the factors that contribute to long-term success, such as investments in research and development or employee wages and benefits. An obsession with shareholder value also turns a blind eye to possibilities for companies to do good for everyone with a stake in their success—not just investors, but also employees, customers, and the broader public.

The consequences can be catastrophic. In April 2019, a federal judge in San Francisco criticized one of California's power providers, PG&E, for paying out $4.5 billion in dividends to shareholders while neglecting routine maintenance, such as trimming trees that might pose a risk to power lines. PG&E's lawyers complained that a thorough campaign of maintenance and inspection would be too expensive, even though the company clearly had capital to spare. California's citizens have borne the brunt of these decisions: Eight people were killed in a 2010 PG&E pipeline explosion, and PG&E equipment was the likely cause of both the 2017 fires in the Bay Area's wine country and the 2018 Camp Fire—the deadliest and most costly wildfire in California's history that killed at least eighty-five people, destroyed more than eighteen thousand buildings, and led to $16.5 billion in damages. Wary that these patterns of neglect would kill more people and damage more property, Judge William Alsup directed PG&E to step up its maintenance work. "A lot of money went to dividends that should've gone to your trees," he said. "Get square with the people of California, who depend on you to do the job safely." But six months later, in late September 2019, PG&E reported that it had finished only 31 percent of the work it had planned for the entire year—leaving Californians to wonder if the dozens of fires that broke out in October 2019 could have been mitigated, if only PG&E had decided to prioritize maintenance over shareholder dividends.[4]

For a company operating in the digital world, this approach might have worked. Google provides many examples of this phenomenon. A Wikipedia page that lists Google products and services includes a "discontinued" section that names more than one hundred that are no longer supported. Readers might remember Picasa, Wave, Dodgeball, Buzz, Aardvark, Health, Knol, Meebo, Orkut, Google+, and so on. Old products were folded into new services, or were so poor to begin with that few people noticed or cared when they were gone, but in other cases users were infuriated by the discontinuation of favorites. Announced to great fanfare, with a flourish of optimism, all of these products ultimately suffered the same fate: decline, neglect, abandonment, and failure.[5]

In the meantime, Google continued to boost its profits—indeed, the company's highly successful strategy for building shareholder value was in part a function of its willingness to jettison some products so it could focus on others. But what's good for Google is not necessarily good for America, and what works for software companies does not necessarily translate to companies that work with physical products and services.

Two other brief examples illustrate how attempts to maximize shareholder value can lead to public detriment. Ironically, these attempts can also harm corporations themselves. The first comes from Boeing, the subject of a bombshell *New York Times* report in April 2019 that documented a "culture that often valued production speed over quality," leading to manufacturing failures in two of Boeing's flagship products, the 737 MAX and the 787 Dreamliner. The evidence published by the *Times* showed how safety and quality concerns were ignored by managers who were keen to preserve the illusion, for executives and shareholders, that deliveries would be made and dividends would be paid—an illusion shattered in late 2018 and early 2019 when two 737 MAX planes crashed, killing 346 people.[6] These poor decisions also cost Boeing dearly: In late 2019, Boeing estimated that the 737 MAX crisis cost the company more than $9 billion and reported a 51 percent drop in quarterly profits.

A second example—a seemingly innocuous reporting of quarterly earnings—highlights how the growth imperative can pit companies against the needs of their customers. On November 1, 2018, Apple reported $62.9 billion in revenue and $14.1 billion in profits for the fourth quarter of its 2018 fiscal year. These figures beat Wall Street estimates and were a significant increase over $52.6 billion in revenues from the previous year. But investors were unhappy, and Apple's stock dropped a whopping 7 percent. Why? In part because sales of its iconic iPhone fell short of investor expectations.

Analysts scrambled to understand why iPhone sales were lower than they projected. At first, Apple CEO Tim Cook blamed declining demand in China, attributable to President Trump's trade war and the slowdown in the Chinese economy. But a couple of months later, Cook acknowledged that the popularity of Apple's program allowing users to replace the battery of their iPhone rather than have to buy a new phone was a factor. This explanation resonated with many iPhone owners: Why would you pay between $750 and $1,100 for a new phone when you could spend $29 to fix the one you already own?

The absurdity of Apple's situation becomes clearer when you think about it in a more holistic way: Apple's stock price dropped because users were—in a very modest way—choosing to fix instead of throw away. But Apple executives and stockholders didn't seem to care about the potential benefits of these choices, such as less pressure to exploit natural resources, or a decrease in waste as perfectly operational iPhones no longer needed to be thrown into dumps and landfills for want of a reliable battery, or freeing up customers to spend their money on something more important. None of these possibilities seemed to matter to Cook, who closed his letter to investors by declaring, "Apple innovates like no other company on earth, and we are not taking our foot off the gas."[7]

If the goal of innovative companies is to constantly increase profit, then anything and everything is fair game for being put in service of that cause—even values like efficiency and sustainability. Despite clear evidence from economic history that we are unlikely to repeat the high rates of innovation and productivity growth of earlier iconic eras in American history, CEOs and managers remain enchanted with innovation and have generated mountains of waste in pursuit of short-term growth. This has done immense damage to individuals and society at large, in the form of burnout, labor exploitation, and environmental degradation—all in service of the elusive and never-ending pursuit of profit.

The core of the problem, then, is the very powerful forces that present growth as a panacea and innovation as its handmaiden. We have already explored one cause of this behavior, which is myopic, short-term thinking manifest in the desire to produce good news for quarterly earnings reports. And it's not pretty when a company that has coasted on high expectations begins to descend into a death spiral.

WHEN THE BUBBLE POPS

The historian David Kirsch is fascinated with an understudied phenomenon in business and technology: failures. In the early 2000s, as pundits and investors were looking for the next digital economy blockbuster, Kirsch was looking backward—at the detritus created by the dot-com crash of 2000–2002. Trained as a historian of entrepreneurship, and the author of a book on electric cars invented in the 1890s, Kirsch wondered what would become of the dot-com era's ephemera: marketing swag, consulting PowerPoint presentations, and ill-fated business plans that had been produced during the exuberance of the boom's peak, from 1995 to 2000. Over time, he came to understand the dot-com mania as an instance of the recurring economic phenomena of exuberance and irrationality—or, in a word, "bubbles."

Kirsch and the economist Brent Goldfarb define "bubbles" as dramatic changes in asset prices that fail to reflect changes in underlying intrinsic value. In other words, bubbles are fundamentally *social* phenomena driven by collective behavior. When people keep telling themselves stories that justify continued faith and investment in a particular market opportunity or way of doing things, a narrative develops and strengthens. These narratives, in turn, help to sustain the collective hallucination.

Bubbles don't lead exclusively to failures. Take the dot-com bubble of 2000–2002, for example. Although companies such as eToys.com, Webvan, and Pets.com did not survive, another character in this story was Amazon.com—arguably the era's most profitable success. Internet businesses and e-commerce were the stuff of a digital gold rush, and extremely high rates of capital "burn" were seen as evidence that a company was following the logic of "Get Big Fast." Reflecting on his loss of $850 million, eToys.com CEO Toby Lenk wrote, "Grow, grow, grow. Grab market share and worry about the rest later."[8] Nearly two hundred Internet-related companies announced IPOs in 1999 and 2000, and investors, for a time, didn't mind that these companies weren't actually making profits. But reality kicked in between March 2000 and September 2002, a period now known as the dot-com crash. When the logic of "Get Big Fast" collapsed, the entire economy felt it: The NASDAQ declined by 76 percent, and the S&P 500 fell 48 percent.

Many companies avoid the extremities of bubbles, but still make big bets on fads and trends with dubious long-term prospects. We started the chapter with General Electric, the lumbering giant that, in the 2010s, followed the cool kids of Silicon Valley when it counseled division managers to "fail fast" and see their company as a "124-year-old start-up." But this was not the first time GE had followed a business trend with disastrous results.

Founded by Thomas Edison and other electrical pioneers, GE was an icon of the twentieth-century American corporate landscape. "We bring good things to life" was a fitting jingle for a company whose world-class research labs and manufacturing facilities anchored communities in Massachusetts, Pennsylvania, and upstate New York. Without question, GE's products improved the lives of Americans: electric lights, radios, televisions, jet engines, medical devices, and much more. Thanks to a steady stream of competent managers and executives, the company became a reliable partner for American consumers, investors, military officials, and even presidents.

Things changed, however, in the 1970s. GE's business slowed. It closed factories and laid off workers. Investigators exposed its decades-long, systematic pollution of the Hudson River with toxic by-products from its factories. It was not yet clear at the time that American companies would not be able to extend the remarkable run of productivity growth that companies like GE had sustained between 1870 and 1970.

When he took over as GE's CEO in 1981, Jack Welch was eager to transform the company and identify new opportunities for growth. One of his early speeches as CEO was titled "Growing Fast in a Slow-Growth Economy," and he took aggressive steps to make his topic a reality. Welch expanded GE into new lines of business, most notably financial services, by acquiring hundreds of companies. In doing so, he set fire to GE's traditional image as a steady and reliable employer. Each year, "Neutron Jack" fired the bottom 10 percent of the company's managers through his so-called rank-and-yank program, an approach that significantly reduced GE's overall workforce.[9]

Through Welch's recasting of the stable manufacturing giant as a financial juggernaut, GE posted some amazing numbers over the course of his twenty-year reign. The company grew its net income from $1.65 billion in 1981 to $12.7 billion in 2000, and "rightsized" its workforce from 404,000 to 313,000. On the stock exchanges, GE's value rose 4,000 percent. The profits came from Welch's decisive turn away from manufacturing (hence the layoffs) and toward financial services. GE Capital's presence in insurance and mortgages, as well as financing for aviation and energy, was, for a few years, well timed to ride the wave of American financial growth.[10]

But the era of fast growth didn't last. When Welch stepped down in 2001, his protégé and successor Jeffrey Immelt took over a company sprinting into a briar patch. Cruel reality hit in 2008, when GE stock plunged from $37.10 to $8.50 per share. The company was effectively saved from disaster by emergency investments, including a $3 billion infusion from Warren Buffett.[11]

As the wreckage of the 2008 financial disaster became clearer,

GE's aggressive move into financial services in the mid-1990s looked worse and worse. It's easy to see why Welch found this strategy appealing—after all, *Fortune* magazine named the energy financing giant Enron as America's "most innovative company" each year from 1995 to 2000. It was obvious that GE needed to find new sources of revenue and new strategies for regaining the confidence of investors.

It's tough not to feel some sympathy for GE, especially in an era when unproven start-ups such as Uber and WeWork were darlings of Wall Street and Silicon Valley investors—while losing millions, if not billions, of dollars every year. In many ways, GE is simply another victim of two factors—economic incentives and cultural fads—that heap rewards on innovators and neglect activities like trimming trees and slow, incremental growth.

Such attitudes are carving a path of destruction throughout American institutions. So far in this chapter we've focused on businesses like GE, PG&E, Google, and eToys.com. But the growth delusion also runs rampant in two large industries that powerfully shape the lives of every American: education and healthcare.

THE COST OF INNOVATION WHERE WE LEARN

Education is a tempting target for disruption precisely because it is so universal—everyone is involved one way or another, and few social institutions have more resources devoted to them. Moreover, there is *always* room for improvement, and plenty of evidence to illustrate how much, and where. On ASCE's 2017 Infrastructure Report Card, American schools earned a barely passing grade, D+. The sorry state of public school facilities was the focus—an area usually beyond the attention of philanthropists and EdTech companies. American public schools, one hundred thousand buildings in total, are home to nearly fifty million K–12 students and six million adults, but the systematic underinvestment in public education has created a $38 billion gap necessary for schools to stay in good condition and provide healthy, safe, and modern learning environments. The report card notes,

"More than half (53%) of public schools need to make investments for repairs, renovations, and modernizations to be considered to be in 'good' condition." But while billionaires flood educators with digital gadgets and promises of "revolution" and "disruption," the report card found that "four in 10 public schools currently do not have a long-term educational facilities plan in place to address operations and maintenance."[12]

Public investment suffered with the 2008 recession. (As recently as 2014, thirty-one states were providing less funding than they did in 2008.) The consequences, for people who care about maintenance and reliability of schools, are discouraging. The report card concludes: "Facing tight budgets, school districts' ability to fund maintenance has been constricted, contributing to the accelerating deterioration of heating, cooling, and lighting systems." The situation leads to a downward spiral of debt and decay, which ultimately will cost school districts more money in the long run.

These problems are not confined to K–12 public schools. Public higher education has also been receiving poor grades, albeit from a different teacher. Each year, Moody's Investors Service publishes a "higher education outlook," and the news in 2018 and 2019 was not good.[13] The core problem was that universities could not meet their revenue growth targets, resulting in the need to control costs.

The existential financial dilemma of higher education in the twenty-first century is simple. The revenue-generating strategies of the late twentieth century—including annual tuition hikes of 5 to 10 percent, which shifted financial burden to students in the form of interest-bearing loans—are increasingly untenable. Administrators are scrambling to keep up with the rising costs of instruction and student services, but traditional veins of cash, such as philanthropic giving and sponsored research, continue to be dominated by a handful of elite institutions. Universities also experiment regularly with "strategic partnerships" with industrial and government actors, often framed in the dialect of innovation-speak—incubators, innovation parks, and so on. But existing evidence suggests that these gambits rarely create the jobs and economic benefits promised by their promoters.[14]

So, what does the euphemism "controlling costs" actually mean in the business of education? Since labor accounts for 65 to 75 percent of costs in higher education, according to Moody's, the most obvious step for school budget directors is to keep salaries flat, avoid hiring more people, and increase reliance on temporary workers and adjunct faculty. But when that's not enough, and administrators keep scouring budgets for other big numbers to cut, their cursors and pencils pause inevitably on maintenance and facilities costs.

The obvious solution would be to increase revenues by increasing enrollments—in other words, to grow their way out of the problem. And thus universities repeat the pattern we have observed in the previous chapter and this one: Leaders facing complex problems revert to chatter about innovation, disruption, and growth. Today's university administrators and faculty face tremendous pressure to create "innovative" programs to capitalize on buzzwords and fads—Big Data! AI! Learn to Code! In the process, they drain resources away from established programs and fundamental subjects such as writing, mathematics, history, and languages that serve all students and generate enduring knowledge and skills.

Teachers, entrepreneurs, reformers, and others in the education industry have followed the lead of practitioners in many other domains of innovation-speak and become increasingly fixated on technological solutions known as educational technology, or EdTech. The computer systems that undergird massive open online courses, or MOOCs, which Clayton Christensen and others hyped, are one classic example of EdTech. Yet EdTech efforts often end in disaster for the schools that adopt the technology and the firms who make it. An article titled "Chronicling the Biggest EdTech Failures of the Last Decade" on the Tech Edvocate website names as an example inBloom, essentially an app store for teachers and a platform for schools to share data about their students—ostensibly to make student data more useful.[15] The company received $100 million in funding, most of it from the Bill & Melinda Gates Foundation, but exploded into nothingness within a year of its launch. Elsewhere, a recent study found that fourth graders who used tablet computers in "all or almost all" classes scored lower on standardized tests than peers who did not.[16]

Despite disappointing educational outcomes and a string of failures, advocates still hold on to the EdTech dream. The scholar Christo Sims did ethnographic research in a New York City school that opened in 2009 and focused on preparing students for a predicted future economy. "The entire curriculum would be designed like a game, and the latest digital technologies would be woven through all classes," Sims wrote.[17] But many efforts at the school fell short, and its programs were never as unique and cutting-edge as its leaders claimed. Sims argues that a kind of techno-idealism keeps EdTech boosters going even in the face of countervailing evidence: Advocates' "collectively lived fictions are maintained, repaired, and renovated despite round after round of often disappointing setbacks." The phenomenon that Sims studied is widespread, as the education writer Audrey Watters has documented and summarized in her sobering panoramic collection, "The 100 Worst Ed-Tech Debacles of the Decade."[18]

Thinking through and planning for maintenance is one area where EdTech efforts repeatedly fall woefully short. In a recent book and related articles, the scholar Morgan Ames examined the One Laptop Per Child (OLPC) program, which came out of MIT's once-vaunted, now-disgraced Media Lab (which fell into disfavor in 2019 after journalists revealed that the lab accepted money from the financier and convicted sex offender Jeffrey Epstein).[19] The OLPC promised to revolutionize education in the Global South by providing cheap and sturdy laptops to individual kids. Ames found, however, that the program's leaders failed to deal with the realities of maintenance and repair. They were giving the devices to *children* after all, who would do God knows what with them. Laptop programs from Thailand to eastern Virginia have similarly foundered on an inability to deal with repair.

In a study of a program that gave students tablet computers at an inner-city school in California, the informatics professor Roderic N. Crooks found that *students* were fixing the broken machines.[20] Technology advocates sometimes argue that it's a good thing if children become the repairers because they learn valuable skills. But Crooks found that the students stepped in because the school system had neither planned nor provided adequate resources for repair. As we've seen, the 2017 ASCE Infrastructure Report Card gave American schools a D+ because buildings are in poor condition due to deferred maintenance. Now we are worsening an already bad situation by loading up schools with hundreds of thousands, perhaps millions, of digital devices and likewise not dealing with their maintenance.

THE COST OF INNOVATION WHEN WE'RE SICK

As we have seen, the growth delusion is the idea that economic growth will solve all of our problems—in both business and our public institutions. The limits of that approach can perhaps be seen most clearly in healthcare, where decay and senescence are inevitable, and faith in never-ending growth or technological fixes is a clear sign of our refusal to confront reality.

At first glance, healthcare provides some of the clearest examples of innovation's benefits. Consider laboratory discoveries such as antibiotics and insulin, which have saved millions of lives around the world, or the vaccines that have lowered global child mortality rates. It is difficult to overstate how the results of innovations in medical science and administration have led to the alleviation of human suffering.[21]

At the same time, the undeniable benefits of innovations in medicine and healthcare have cast a harsh light on the problems that remain. Healthcare is at the heart of a paradox that has aggravated generations of American policy makers, reformers, and health professionals: Americans spend the most money per person on healthcare—up to twice as much as citizens of other high-income countries, according to some studies—but with worse outcomes in areas such as infant mortality and life expectancy.[22]

The pressure to find solutions has created a system that is horribly misshapen and full of contradictions. For example, Americans are global leaders in digital technologies, but health maintenance organizations still rely on paper records, often distributed by fax machines or hand. Another contradiction is visible in the phenomenon of so-called orphan diseases, conditions that affect less than two hundred thousand people nationwide. Some examples will sound familiar, such as cystic fibrosis or Lou Gehrig's disease, but others, with patient populations below one hundred, are virtually unknown. Since market mechanisms tend to ignore orphan diseases—the costs of research and treatment are too high, given the number of patients who will pay for them—Congress, the National Institutes of Health, and the Food and Drug Administration (FDA) created financial incentives to encourage researchers and drug companies to work on them. But this approach is insufficient: Congress passed the Orphan Drug Act in 1983, and although nearly eight hundred orphan drugs have been approved by the FDA since 1983, no drugs have been approved for 95 percent of rare diseases.[23]

Each of these problems flows from a common source: Americans aren't putting their good ideas to work in a systematic way that benefits all of their fellow citizens. And that general problem fits the familiar trend we've seen in business and education, where leaders choose to steer their organizations toward innovation, with the implicit assumption that doing so will lead inevitably to financial success.

A clear example of the Innovation Delusion in healthcare comes from the field of organ replacement. Successful outcomes for organ transplants, despite all of the potential risks and sources of failure, have steadily improved over the past seventy years since the first kidney transplant at Chicago's Little Company of Mary Hospital. But advances in the specialty took a dramatic turn in the 1990s, when medical researchers began to experiment with embryonic stem cells for regenerative therapies. They saw the potential for stem cells to differentiate into any and all types of human cells—which could, eventually, produce artificial organs, targeted cellular therapies, and tissues that could be engineered to repair damaged organs. The allure of these breakthroughs was too much for funders and researchers to resist. Medical journals and top research facilities, such as the Mayo Clinic, are positively abuzz about the potential for innovation. The National Institutes of Health created the Regenerative Medicine Innovation Project in 2016, putting $30 million each year toward accelerating discoveries in the field.

While they wait for these investments to pay off, experts worry that limited regulatory oversight is creating opportunities for unscrupulous entrepreneurs to take advantage of patients with unsafe products and treatments. Champions of innovation will reliably tell you that stifling regulations are their biggest obstacles—but when it comes to biomedical treatments like growing new organs, caution seems warranted. In the meantime, resources and expertise are being diverted from safer approaches for organ transplants: According to the American Transplant Foundation, almost 108,000 people are on waiting lists for a lifesaving transplant, and an average of 14 people die every day while they wait. The Department of Health and Human Services reported in 2015 that the number of donors has remained stagnant, and the number of organs recovered from living donors dropped 16 percent over the past decade. Physicians are sometimes reluctant to speak publicly about the dilemma. One told us that the transplant unit in his hospital is decades old and "suffers from inadequate infrastructure and limited nursing resources." He's frustrated that there are billions of dollars going into regenerative medicine start-ups that he sees "going nowhere," but he also feels like he has no other choice but to apply for research funding in the same field.

The same perverse incentives are evident when we consider the American approach to elder care and gerontology, the study of aging. In *Being Mortal,* Atul Gawande describes how effective gerontological treatment can be for the quality of life for elderly patients. A great deal of gerontological work focuses on maintaining bodies, or more accurately, helping other people maintain their own. Gawande writes lyrically about a doctor looking over an elderly patient's toenails, checking to see if she is taking care of herself. The pain that comes from ingrown toenails can be easily avoided, but people lose flexibility and vision as they age. Untrimmed toenails aren't fatal, but they are signs that a patient could use more attention and support. Yet Gawande notes that the field of gerontology is dying in many parts of the United States because young people choose to go into specializations that are seen as more innovative and cutting-edge, and thus receive higher pay.

Gawande is one of many medical experts who are warning about the crisis in elder care. Federal data confirm what many American families already knew: Most nursing homes do not have adequate staffing levels. The problem is only getting bigger and more pronounced as baby boomers age and life expectancies increase—but the prospects for acceptable standards and quality of life will not be equally distributed. Unfortunately, this situation is a predictable outcome of a healthcare system that rewards innovation, undervalues maintenance and care, and has weak mechanisms for ensuring equity across all populations. These are the consequences of a culture obsessed with superficial delusions of growth as a panacea—and blind to the profound damage to human health, learning, and livelihoods.

We chose these few examples of how the Innovation Delusion permeates American business, education, and healthcare from dozens that we have witnessed personally, and hundreds more that have been suggested to us. We have no doubt that readers will think of many more. There is a clear, common theme for all: The Innovation Delusion degrades our societal infrastructure, undermines its public and private institutions, and harms our health. In the next chapters we will see the profound consequences it has both for our private lives and for the value we assign to the different kinds of work that sustain our society.

CHAPTER SIX

The Maintainer Caste

ON THE STATUS WE GIVE DIFFERENT KINDS OF WORK

Ralph works in the IT department at a university in the American Midwest. Tall and solidly built with graying hair and a beard, he has never cared much about style and is rarely seen in anything other than jeans and a T-shirt. Ralph grew up in an impoverished Rust Belt town near Chicago, long after local steel mills had closed down. He studied physics in college and moved between several jobs after he graduated, ultimately landing in IT. All of Ralph's computer skills are self-taught, yet he managed to crawl his way up from an entry level job at $12 an hour to a salaried position that pays more than $60,000 a year.

One of Ralph's favorite things about his job is his IT colleagues. He says they share a "How can I help you?" attitude. Few if any of them are showboats looking for personal glory. The IT workers face constant sets of frustrations. "A refrain you hear around my job all the time is 'Shitty software being shitty,'" Ralph told us. Yet camaraderie makes the work worthwhile.

The IT crowd's spirit of humility and generosity is sometimes harder to find in the people they serve, who demand that problems be fixed immediately and assume that IT workers are the ones to blame.

Ralph describes users who contact his office immediately when they get an error code and say that they don't know what's going on, and that the IT department must have done something to cause it. This happens even when the error code explicitly tells them that the problem is a result of the user's actions, like editing core computer files.

Ironically, Ralph gets some of his most head-scratching requests from the computer science department. The professors in that field, who are tasked with pursuing "innovation" in various ways, often ask the IT department to do things that are physically impossible, such as making information move over networks faster than it possibly can—and they often do so with a condescending tone of entitlement.

A professor once complained that his system was running too slowly. When the IT team investigated, they found that he was running programs over a network connecting multiple computers, as opposed to on a single machine. Sending information from point A to point B took time and computing overhead because the system obeyed the laws of physics. Working this way, as Ralph put it, "is flat-out never going to be as fast as a single-disk local" run on the professor's own computer, but the professor didn't seem to understand that.

Part of the issue is that there is a big gap between academic theory and realities of running a bunch of actual computers. Ralph puts it a little more bluntly: "Computer science isn't science, and a good chunk of [the professors] sure as shit don't know how a computer works." When the IT department gets help tickets asking it to do the physically impossible, he says, "You have to write back in the politest way you can possibly tell someone that they are fucking ignorant." The way he sees it, the professors' expectations are formed in theory land, but they don't bother to think about how to make their ideas work in the real world.

Ralph plays an important role at the university. He keeps 450 physical Linux machines running—equipment often used by a bunch of rowdy undergraduates—as well as a number of virtual computers. Yet it will surprise no one that the professors are the ones with status at the college. The university website is festooned with innovation-speak, including news items on how professors have introduced this or that innovation, and how the school held hackathons, coding camps, and other events meant to turn students into disrupters. The people who keep all of the computers running in the school are, of course, nowhere to be seen on the webpage. Though their labor is crucial, the IT workers are overlooked and taken for granted.

Ralph's experience is not unique. Within organizations and society at large, maintenance roles often fall at the bottom of status hierarchies. Nearly all maintainers experience condescension on the job, whether it takes the form of being ignored, talked down to, or taken advantage of. In many organizations, for instance, janitors and maintenance workers are required to wear uniforms—often one-piece coveralls—that mark them out as maintainers. Where do these traditions and mindsets come from? In this chapter, we'll draw on social science and interviews to suggest that maintainers fall into a caste-like social hierarchy—a system that we must rethink in order to make our society more sensible and just.

WHAT OUR PARENTS TEACH US ABOUT MAINTAINERS

The division of labor is older than human civilization. In many non-human animal species, such as ants and bees, different types of individuals—workers, drones, queens, etc.—have different roles. One study of harvester ants (*Pogonomyrmex barbatus*) found that different groups of workers were responsible for one of four tasks: foraging, patrolling, nest maintenance, and the upkeep of the colony refuse pile.[1] Note that two of those four tasks are explicitly maintenance-focused.

Humans also divided labor very early on. Hunter-gatherer societies were, and still are, marked by a sexual division of labor, with men and women focusing on different tasks. But the unequal division of maintenance and cleaning work accelerated as societies became more complex and stratified. In ancient Greece, for instance, Plato argued that the most important thing for philosophers was *skhole,* leisure time spent in the pursuit of knowledge, a word that eventually came down to us as "school." What made this free time possible for philosophers and other Greek navel-gazers was the work of slaves and servants who kept society running. Over time, in Western culture, these realities gave birth to the distinction between work of the "head" and work of the "hand": Mental labor was seen as belonging to a higher order than manual—or "menial"—labor, which was done by individuals in the lower ranks.

Status hierarchies around work, including maintenance and repair, become most visible in societies with explicit caste systems. India has the most famous caste system in the world (though experts debate how old it is and where it came from). In the Indian caste system, families of Dalits, or "untouchables," do work known as manual scavenging, which involves cleaning open-air toilets and sewers by hand. Ninety percent of this dirty and dangerous work is done by women.[2] Manual scavengers frequently die from suffocating on poisonous gases.

India is far from the only society with a caste system, however. In Yemen, al-Akhdam (which literally means "the servants") are a minority group who are socially segregated from the rest of society and only able to do low-status, dirty work like street sweeping and toilet cleaning. A Yemeni proverb hints at their place in society, "Clean your plate if it is touched by a dog, but break it if it's touched by a Khadem."[3] Caste and caste-like systems exist in many other nations around the world.

We shouldn't kid ourselves that America and other Western societies are free of such hierarchies. They aren't. Historically, of course, the United States was a slave society, and black slaves did wide varieties of maintenance work, including keeping house. But the end of slavery did not entail the end of social hierarchies or occupational status systems. In Virginia and other southern states, for example, a number of jobs haven't been covered by the state's minimum wage rules, including hotel doormen and domestic help—jobs traditionally held by African Americans.

Since the 1920s, social scientists have been studying how given jobs in the United States fit within social status—something called occupational prestige. In one oft-repeated study, researchers asked individuals to rank a list of twenty-five occupations in order of social status. The rankings turned out to be remarkably stable, changing little even as the study was repeated over several decades and conducted across the United States. Bankers, physicians, and lawyers—"head" workers—always landed at the top; janitors, bricklayers, and ditchdiggers were always at the bottom. Maintainer roles, like plumbers and barbers, reliably fell in the lower half of the rankings. Many of these jobs are crucial—society could not go on without them—yet people usually prefer to keep them out of sight and mind. As one group of social scientists put it, "Despite its fundamental nature, dirty work seldom makes it into 'polite' public conversation. There is an apparent understood obligation to protect society from its dirty work, and this stigmatizes (taints) its dirty workers."[4]

The prestige of some jobs, especially those connected to technologies, *does* change quite a bit over time, however. For example, earlier in U.S. history mechanics and electricians were fairly high-status roles, positions that working-class men who wanted to get ahead in life aspired to. Over time, though, as the historian Kevin Borg has shown, the job of being an auto mechanic became a vocation for "unpromising" individuals who were not "college material."[5]

There's every reason to believe that the downward trajectory of occupational prestige will continue with the technology-centered jobs that are popular today. *Wired* magazine set the Internet on fire when it published "The Next Big Blue-Collar Job Is Coding," in which the technology journalist Clive Thompson argued that many IT jobs are akin to blue-collar jobs, and training for such work will likely become more vocational over time. Even today, as we see elsewhere in this book, most IT jobs are focused on maintenance. What Thompson didn't emphasize in his article was that this process will almost inevitably lead to declining wages and compensation in the IT field, as more and more workers enter the market.

While some employment is remarkably stable in the status echelon, change is possible over time. But if our ranking of occupations is reliable—in the sense that we all tend to think the same thing—where does our intuition of what job fits where in the hierarchy come from? Well, from lots of places, but perhaps, most important, we are *taught* how to stratify work.

In a classic article titled "What Do Animals Do All Day?" the sociologist John Levi Martin created a database around Richard Scarry's children's book *What Do People Do All Day?*[6] In Scarry's books about Busytown, children are introduced to occupations, with different animals doing different jobs. Martin did a statistical analysis of the relationship between the animals in Scarry's book and the jobs they're depicted performing. Martin found a correlation between animal species and statuses. If you live in Busytown, you want to be a predator. The mayor and the airline pilot are foxes, and the doctor is a lion who drives a Land Rover. What animal does the most blue-collar work? The lowly pig, which Martin describes as standing in for "America's Working Man." Not only do pigs in the book do stigmatized manual labor, including jobs as sanitation workers, they also tend to cause many of the mishaps and accidents in Busytown. Scarry's character Mr. Frumble—whose primary narrative role seems to be to cause accidents through foolishness—is a pig.

Martin's article is fun to read and funny in parts, but his point is serious. As he writes, in reading Busytown, "the child learns not just what people do all day, but what *kinds* of people do what *kinds* of things." In other words, when parents spend sweet quality time reading Busytown books to their attentive children, they teach their kids about occupational stratification. As Martin points out, it's not clear whether Scarry produced this occupational hierarchy consciously or unconsciously. He may have just been unthinkingly representing ideas about status that *he* was taught as a kid.

We get ideas about what kinds of people belong in what kinds of jobs from other places, too. A friend of ours who lives in a Manhattan high-rise has recounted how she's had to talk to her young daughter about race in the United States because all of the African Americans the young girl regularly encounters are doormen and cleaning ladies who belong to the class of service workers. If left unchallenged, the child's daily experiences might suggest that African Americans *should* be in only these roles.

None of this suggests that the same kinds of individuals occupy the same jobs in every location, even if we just focus on the United States. Both of us have lived the itinerant lives of academics and moved around quite a bit from our teens into our late thirties. In rural areas of Illinois, New York, and Virginia, we've noticed that "low-skill" jobs, like janitors and fast-food workers, are performed primarily by lower- to middle-class whites—in the case of fast-food workers, sometimes teenagers. In urban centers like Chicago and New York City, these jobs are often held by racial and ethnic minorities, frequently recent immigrants. *What doesn't change* between locales, however, is the relative status of the roles themselves. It's not like the janitors in Arcola, Illinois, or some other midwestern farm town all drive gold-plated luxury vehicles while the medical doctors putter around in ancient used rust buckets that are falling apart.

Social stratification in the job market is real, and certainly by the time we are teenagers we have a good sense of where a given occupation fits in the hierarchy. Learning this is a part of growing up; this knowledge is a basic competence of existing in a society. Again, we gain much of this knowledge from parents and other loved ones as well as from direct experience of our social world. But once we enter school, and especially college, we learn lessons that drive many of these ideas home.

WHAT WE LEARN AT SCHOOL

As the ideology of innovation-speak permeates every aspect of our culture, it has hit our education system—from pre-K to doctoral programs—particularly hard and reinforced the status differences between innovators and maintainers. In the United States, innovation-speak is often tied to STEM education—science, technology, engineering, and mathematics. Producing STEM students is held up as one way to sustain and increase innovation at the national level while helping kids get good jobs when they grow up. Students are encouraged to attend hackathons, coding camps, robot clubs, and other extracurriculars aimed at fostering their innovative potential. Pedagogical fads like Design Thinking K12 (DTK12) and new university centers like James Madison University's X-Labs claim to develop students' "innovation capabilities."

In chapter 2, we questioned the idea that innovation comes from "innovators" who have some special "capabilities" that can be taught. There is no single set of capabilities or skills that links together the inventor Thomas Edison; the creator of nylon, Wallace Carothers; and entrepreneurs like Oprah Winfrey and Arianna Huffington. Some innovations come from gregarious extroverts who see opportunities in every social situation; others from shy autistics who would rather push pins under their fingernails than hang out at a party. Moreover, there's plenty of evidence to doubt the idea that we can teach general "critical thinking skills," as James Madison University's X-Labs claims to do.[7] Most innovations are incremental and come from individuals who are subject matter experts and who apply themselves to improving the domain they know so well. There is no shortcut to decades of training and hard work.

But beyond pushing shaky educational theories, the real damage stems from how universities lionize innovation when most of their students end up doing non-innovative but completely crucial work in our society. Over the last few years, we've given several talks about The Maintainers to classrooms full of college students. When we ask how many of them want to graduate to become mechanics, electricians, IT support staff, or do other jobs centered on maintenance and care, no hands go up. Our question is a joke, of course. People generally don't go to college to enter "trades." But our point is to make students think more broadly about their aspirations. Students assume they will end up in innovator roles simply because they've been told they will.

From 2012 to 2016, both of us worked at Stevens Institute of Technology, which had (somewhat awkwardly for us) trademarked the motto "The Innovation University." As part of their senior engineering capstone project presentations, Stevens students were required to describe how their projects were innovative. Of course, most of the projects were not in the slightest innovative, so the primary lesson students learned was *how to bullshit* and sell themselves as something they were not. Performance and drawing attention to one's work as novel is an important part of being an "innovator," something we'll return to in a moment. But the deeper problem was how badly the Stevens innovation requirement misconstrued the nature of engineering. The sheer reality is that about 70 percent of engineers maintain and oversee existing systems.[8] Only a small minority of working engineers have jobs focused on invention and the "research" part of R&D. As a rule engineers are maintainers and operators, not innovators.

The same goes for computer science, the current hot field at many universities. Data show that somewhere between 60 and 80 percent of software budgets are spent on maintenance.[9] And we haven't even counted the college grads who enter non-software areas of computing, like IT infrastructure, user services, and network engineering. Most graduates will end up working as maintainers. Given that universities are places where people are supposed to chase truth, we believe schools would be much better off putting forward a realistic picture of the world and what their students do in it. They might find ways to highlight maintenance thinking and the value of the maintenance and operations work that most students will wind up doing.

An even deeper myth, linked to innovation-speak in several ways, is the idea that you must go to college to achieve a middle-class existence. While it is true that college graduates tend to make more than people who work in the trades, focusing on this fact alone neglects other factors. For instance, trade school costs an average of $33,000, whereas bachelor's degrees cost about $127,000, if you include tuition, living expenses, and interest on student loans.[10] Moreover, about 70 percent of students take out loans to pay for college with 20 percent of them borrowing more than $50,000. Paying off these loans and interest cuts into their earnings, often for decades. Furthermore, trade school programs typically last only two years, which gives their graduates at least a two-year head start on earning income over their peers who graduate from four-year colleges—even more when you consider that it takes many students more than four years to finish their college degrees, provided they don't drop out altogether.

This myth that college is the only route to a decent living isn't uniquely American either. Melinda Hodkiewicz, an engineering professor at the University of Western Australia, has worked for decades on the topic of maintenance, both as a practicing engineer and as an academic researcher. As Hodkiewicz explained it to us, changes in Australian university policy in the early 2010s made it easier for students to pursue engineering majors instead of vocational training. There were several unintended fallouts: The new students were often poorly equipped for rigorous math and science courses, and there is a pressing need for people in the trades. As Hodkiewicz put it to us, "We now have a generation of unemployed and—some would say—unemployable engineering grads who wouldn't dream of taking a technician job."[11]

People go overboard sometimes and argue that we need more tradespeople in the United States. Within the last decade, there has been much debate/talk/discussion about a so-called skills gap, a significant mismatch between the needs of industry and the supply/availability of skilled individuals like welders and electricians, who are trained in trades. Recent studies suggest that the skills gap does not exist, though experts will likely continue to dispute the matter for some time.[12] (On the other hand, where we live, in southwest Virginia and central New York, it is very difficult to find people to work on your home, including exterior work, like cleaning and staining wood houses, and interior work, like plumbing. The problem is well known and discussed by all, including people in the trades. So it appears that geographical pockets do exist where more tradespeople would be useful.)

Our argument is a different one, though. The myth of the necessary college education encourages us to think that the job that pays the most money is somehow the *best* job for us, when in reality people find more joy, meaning, and pleasure doing work that suits them. It also reinforces the ideology that maintenance-centered trades are somehow "beneath" us and unworthy of aspiration.

One of our students faced this crisis personally some years back. The young man was pursuing an (extremely expensive) engineering degree, but he spent much of his time working for a heating, ventilation, and air-conditioning (HVAC) company that did maintenance, repair, and new installations. He *loved* the work. The physical labor was rewarding, he said, and he found great joy in "working with the guys." The teamwork and mutual recognition made the job much more engaging than his schooling and other jobs he'd worked. Adding to the irony, he was well paid—so much so that he didn't even care to make more money—and it appeared that he stood poised to control the company someday, in which case he would have easily made six figures a year.

Why, then, Lee asked, was he torturing himself by taking classes that did not interest him, getting a degree he would not use, and adding to his mounting student loan debt? The student said his mom and dad didn't want him to drop out. They saw college as the key to his future, even though he knew it was not. While we certainly understand that parents want the best for their kids, in this case their insistence that their son stay in school was clearly hurting him, emotionally and physically, as he worked a demanding job on top of a full class load.

An obsession with innovation infects the lessons children learn about work and status, and twists the experiences of students in higher education—ever toward innovation and away from reality. This spin continues once we enter the working world, where misguided managers chase innovation and encourage us to disregard the maintainers.

WHAT WE LEARN AT WORK

The associate dean of libraries was talking about "innovation" again, tossing around terms like "digital humanities," "digital transformation," and "virtual reality." The library staff had grown numb to these speeches, in no small part because the talks were usually accompanied by a lack of follow-through. The associate dean would get all hot and bothered when some new project came around, only to stop giving it attention and resources when he got bored and hopped onto the next new thing. The library had completed three strategic plans in about as many years. The staff eventually realized that *performing* being innovative was the way to reach their boss and started keeping a lexicon of terms that would be more likely to sway him. Wouldn't your project be better with *virtual reality*? they would suggest. Meanwhile, the work of keeping the library going and providing services was often ignored.

This library tale—an amalgam of stories we've heard from professional librarians around the country—highlights how maintenance work can be overlooked and under-resourced in professional settings. Soon after we started talking about maintainers, people started telling us about Susan Cain's book *Quiet: The Power of Introverts in a World That Can't Stop Talking*. In *Quiet,* Cain argues that our culture overlooks and undervalues introverts, individuals who prefer to work alone and are reserved and quiet in social situations. Classic self-help texts like Dale Carnegie's *How to Win Friends and Influence People* are basically primers on extroverted behavior. Such outgoing behavior is prized and rewarded in organizations and society at large, whereas introverts often find it hard to be heard or recognized.

We see two big connections between *Quiet* and what we've been hearing within The Maintainers community: First, like introverts, maintainers often work quietly in the background, keeping things chugging along while "innovators" get the glory. Our society tends to ignore such people and neither recognizes nor rewards them, which creates all kinds of problems, both for the maintainers and for the society itself. For instance, if employees at an organization find it hard to get credit and promotions for maintaining open-source software—something we've heard a lot—they become frustrated, even resentful, and therefore suffer personally. They are also more likely to move on to a different job and be replaced by someone with less experience. In other words, often enough, the *software suffers, too*—and so do the users who depend on it.

Second, we and others have found that maintainers often (but not always) *are* introverts. They prefer to work alone and find extended social interaction stressful and unpleasant. Ralph, the IT worker we met at the beginning of this chapter, said that he enjoys working with his other IT peers because they "aren't showboats," and they "genuinely enjoy helping people with their problems." The flip side of this introversion, however, is that maintainers can find it hard to advocate for themselves and their labors.

Humor is typically used by the popular media to depict maintainers who are out of sight and out of mind in businesses. In the British sitcom *The IT Crowd,* two men—their boss calls them "standard nerds"—toil away in a dirty and messy basement office, resolving computer problems for the offices above. They answer the phone with the question "Have you tried turning it off and on again?" The two men complain together in the first episode, "They have no respect for us up there. No respect for us whatsoever. . . . It's like they're pally-wally when there's a problem with their printer, but once it's fixed, they toss us away like yesterday's jam."

In the early 1990s, the journalist Lesley Hazleton witnessed something similar when she decided to learn more about cars by working as a mechanic. Broken things have a kind of magical power, in that they invert social hierarchies: The job of auto mechanic is usually considered to be fairly low in status, but when a car breaks, the power dynamic shifts. According to the historian Kevin Borg in *Auto Mechanics,* Hazleton "recounted the time a doctor brought his BMW 535 in for a new exhaust system. Placed in the awkward position of having to wait in the shop while the mechanics worked on his car, the doctor tried to ingratiate himself by telling off-color jokes and complaining that he really did not make that much money, perhaps no more than they did. Yet once the car was finished the doctor flipped the mechanic a twenty-dollar tip, hopped into his repaired status symbol, and sped away."[13] Some people feel the need to curry favor with and act humanely toward low-social status maintenance workers only when they need their help. Once the work is completed or if they don't need work done in the first place, they simply ignore maintainers.

In Reddit's Tales from Tech Support, IT workers relate stories of how users—often other employees within the same company—misunderstand basic prompts, make condescending assumptions that the IT staff is at fault, and sometimes even scream at them. When word got out that we were working on this book, we would receive anonymous messages through friends and interviewees, including this one from a contact who worked in IT: "So many problems come to IT which are not really IT problems, such as bad design, bad people, or bad choices."

One IT worker we interviewed, whom we'll call Tom, previously had a job at a software company in the Midwest known for innovative products that were popular with universities. Tom's job was to support the hardware that controlled the company's sales, human resources, and other functions. The company's executives and upper management were obsessed with the system's new software features, but the hardware got little attention or resources. "Everything was held together by duct tape and baling wire," Tom told us. The systems were so old that the equipment manufacturers no longer supported them, and Tom and his co-workers had to compensate by buying spare parts on eBay. It was post-2010, but some systems were still running Windows 95.

Tom also told us about a server farm the company ran at a different location in the same town. Alerts warning that the server farm was too hot were constantly coming in. Instead of working to improve the HVAC system or changing the ventilation ductwork, the company "wrapped the back of server cages where the heat would come out, removed some of the drop tiles in the ceiling, and just pointed some floor fans up behind the server racks to vent the hot air upwards," Tom explained.

"The jokes write themselves." Tom groaned. "We'd say, 'If anything ever happens to the servers, at least they're already in trash bags.' It was awful—lack of planning, lack of sufficient knowledge of how to actually run infrastructure. It was funny but it was also depressing because I had to deal with it."

Just as with housework, which we will examine in the next chapter, gender plays a big role in who is assigned maintenance work. Librarians often complain to us that women are more likely to be tasked with maintaining existing programs than new initiatives. Studies have also shown that women do more so-called office housework and other types of tasks that do not lead to promotions. Over the long haul, this discrepancy can result in real differences in wage growth and a lack of diversity in the type of person who reaches the upper echelons of an organization.[14] Being a maintainer comes with real social and economic costs.

THE HIGH COSTS OF LOW STATUS

If the egos of those who perform maintenance were the only thing at stake, the problem would hardly be worth discussing. Lots of things are frustrating in our modern, bureaucratized lives. Welcome to the world. But we think the personal costs of maintenance work being considered low status go much deeper and take at least two forms: a lack of recognition and a lack of compensation. A third issue is that maintainers often aren't given enough resources to do their work, as Tom's story and the accounts of librarians highlight.

We believe that recognition and dignity are basic human needs, and this view is supported by a long tradition, going back at least to the philosopher Georg Wilhelm Friedrich Hegel (1770–1831), if not earlier. Indeed, much of what is sometimes called "identity politics" today is fundamentally about recognition. More recently, the renowned Canadian philosopher Charles Taylor has argued that recognition—the public acknowledgment of an individual's worth—is a basic need and right and that "nonrecognition or misrecognition can inflict harm, can be a form of oppression, imprisoning someone in a false, distorted, and reduced mode of being."[15]

Since starting this project, we've heard from many people about the frustrations of doing maintenance work that is undervalued, and the costs of this frustration have been backed up by academic research. In one study, researchers surveyed 199 building cleaners at a large public university and did in-depth interviews with twelve of them.[16] The researchers found that "cleaners experienced invisibility at work (not being recognized or acknowledged by customers) and invisibility of work (feeling that work is ignored or unappreciated)." One cleaner described work's greatest heartache as "how people can pass you by as if you're invisible." Others reported not receiving a "hello" or "thank you" when entering a room; feeling "like a shadow" or a ghost, like people "look right through you"; or being treated like a nonhuman object below contempt.

One interviewee told the researchers, "There's some weird moments where a dude will fart around me, that makes me feel like, 'Yeah, you really don't care about my opinion at all.' Then there's a half of me that's laughing about it. When a dude's like coughing up a loogie [mucus], that gets like, 'Wow, dude, I'm here. You're not even embarrassed that I'm here. I'm that little.'" As one janitor put it, "They really don't want to see us." Among other things, these experiences led to feelings of resentment, to perceiving, as the researchers summarized, "students and faculty to be entitled and dismissive." One cleaner griped to the researchers about "ignorant people who don't acknowledge your presence because they think you're beneath them." While these stories lie on the extreme end of what maintainers experience on the job, such mistreatment can lead to emotional and psychological suffering over time.

The foregoing reflections can make it seem like managers and human resource officers could solve the problems of maintainers with employee-of-the-month programs or Celebrate Maintainers! company picnics. But we don't think this is enough. Many maintainers are paid so little they cannot afford a stable middle-class life. Focusing too much on recognition in the face of economic hardship can be perverse, as Nancy Fraser, Lois McNay, and other feminist thinkers have pointed out.

Many American families struggle to make do. Over the last few years, we've closely followed the work led by Stephanie Hoopes, the director of the ALICE project at the United Way. ALICE stands for Asset Limited, Income Constrained, Employed. It's another way of talking about the "working poor," a term Hoopes and her colleagues choose not to use, because poverty should not be stigmatized, and no one should be ashamed of struggling, especially given how many families find themselves in that position.

Hoopes began her career as an academic, earning a PhD in political economy from the London School of Economics and teaching at the Universities of Sussex and Birmingham before landing at Rutgers University–Newark. But increasingly, her work has come to focus on examining and understanding economic struggles in the United States. A study of a low-income community in Morris County, New Jersey, gave birth to the ALICE project and changed Hoopes's life.

Hoopes and her colleagues created ALICE to address this basic problem: Soon after President Lyndon Baines Johnson announced his War on Poverty in 1964, the (formally defined) federal poverty level has become the standard way in which policy makers and public figures talk about being poor in the United States. Yet, many critics have pointed out flaws in the FPL—most important, that the FPL does not account for inflation or the current cost of living. In 2019, for instance, the FPL for the forty-eight contiguous U.S. states and the District of Colombia for a two-person household was $16,910.[17] It is hard to imagine two people living anywhere on that much money, let alone in a city like New York or Washington, D.C.

Hoopes took a different approach. She and her team created a new measure, the Household Survival Budget, which estimates "the total cost of household essentials—housing, child care, food, transportation, technology, and health care, plus taxes and a 10 percent contingency." The ALICE team creates this estimate at the county level, recognizing that these things vary *a lot* depending on where you live.

ALICE's findings can be shocking. The official U.S. poverty rate was 12.3 percent in 2017.[18] Using the ALICE threshold, Hoopes found that closer to 40 percent of American households were struggling to make ends meet. Geography made a huge difference. In Alabama, for example, 43 percent of households were ALICE, but county averages varied from 27 percent in Shelby, which contains some relatively affluent suburbs of Birmingham, to a staggering 71 percent in rural Perry.

The original ALICE report on Morris County, New Jersey, received media attention, and elected officials and advocates started using its numbers when talking about economic hardship in the state. Hoopes decided to leave her academic job to work on ALICE full time. To date, the ALICE project has released reports on eighteen states and published other research. The ALICE team does not make policy recommendations; rather, it seeks to draw attention to the issue of economic hardship and provide facts that do a better job than other measures, like the federal poverty level, in describing the struggles families face.

A few years ago, Hoopes reached out to us because she noticed that the heads of many ALICE households worked as maintainers. It was fortuitous for us because we were struggling with the same realization: Many (again, not all) maintainers were poor, or on the edge of poverty, even though they worked as hard as anyone else.

We conducted a thought experiment with Hoopes. We divided the Bureau of Labor Statistics' occupational categories into two buckets: innovator jobs (with two subcategories: inventors and adopters) and maintainer jobs (with two subcategories: nurturers and infrastructurers—the latter being our term for people who maintain physical things like roads and computers). This exercise was just enough to give us a general sense of things. And we didn't dispute the professions' images of themselves: for example, we counted engineers as innovators even though we know that the work of most engineers has almost nothing to do with innovation.

Perhaps unsurprisingly, the *vast majority* of people work in maintenance occupations—about 95 percent of people were maintainers. But just as important, most of the people who fit the criteria of ALICE worked in maintainer roles. Put another way, while not all maintainers are ALICE, most ALICE families are headed by maintainers. We found that about 64 percent of U.S. workers fell within the infrastructurer category, and 65 percent of infrastructurers made less than $20 per hour, or $40,000 a year for a full-time, year-round job. Many of these workers live paycheck to paycheck and struggle to cover basic necessities. Because they are unable to finance even a basic household budget, they have no surplus for savings, retirement, or investment in education, and they are more likely to require public assistance.

In popular and scholarly discourse, many of these jobs are referred to as "low skill"—the idea being that because they do not require special skills, more people can do them and therefore the labor market "naturally" sets wages low. We reject this way of talking. First, these workers *do* have skills that others do not have, including physical endurance not shared by the kinds of soft, pasty, callus-free people who like to talk about "low-skill jobs." Often, talk of skills is actually talk about social status and little more. Second, while chatter about "low-skill jobs" often leads people to assume we need to educate workers so they get "skills" and thus more lucrative employment, this is a fantasy, and a sick one at that. *Someone* has to do this work, and many people who end up in these positions will stay in them for most of their lives. Given the necessity of this work, our goal should be to ensure that individuals can make a decent living and support their families while doing it.

Picture the job market as a kind of pyramid, with the very few innovators sitting at the top being supported by a broad base of maintainers. This pyramid has several implications for public policy. In chapter 10, we will examine some ways that we might improve the lives of maintainers. But for now, we should set aside what *won't* work. Defenders of innovation policy would say that government action around innovation and entrepreneurship can help create jobs, even for maintainers. But the pyramid makes clear that there is simply no reason to believe that innovation is going to create enough jobs or economic growth to address the problems many maintainers face. If 95 percent of workers do maintenance-type labor and 60 percent of those people are ALICE or close to it, how is innovation supposed to change this picture?

The truth is there isn't enough technological change or economic growth that will fundamentally shift the picture we see in the pyramid. While billions of dollars of public money have been invested in research conducted by fledgling technologies like nanotechnology and biotechnology—which boosters claimed would create booming new industries and thus jobs and economic growth—the outcomes have rarely lived up to the hype. Sure, we can continue to pour money into R&D and other innovation strategies, as long as we are realistic about what benefits will (and won't) flow from our outlay of capital. (Reasonable friends of ours believe that U.S. innovation policy is actually more about global geopolitics, especially conflict with China, than it is about economic growth.) But to address the cold, hard realities faced by ALICE workers, we are going to need other tools and other ways of thinking.

CHAPTER SEVEN

A Crisis of Care

MAINTENANCE IN OUR PRIVATE LIVES

Did you know that there aren't just one but *three* self-help books that seek to apply the concept of "disruptive innovation" to the human soul? We bought the one by Whitney Johnson, *Disrupt Yourself,* wondering if it would be full of hilarious stories of, for example, eating suspicious street food or large platters of greasy grub at 3:00 A.M. after a long, rowdy night of drinking with old friends. In our experience, this is a tried-and-true method for profoundly disrupting your intestinal health—not to mention whatever you had planned for the next day. Sadly, *Disrupt Yourself* ended up being a series of self-help truisms on changing your life and advancing your career, all dressed up in innovation-speak.

There are lots of silly self-help books out there, but *Disrupt Yourself* is part of a much larger phenomenon. Entire industries, from exercise apps to dieting consultants, promise to transform our lives and make us into new people, even though time and again these schemes turn out not to work. You may hanker to change your life and escape the negative repetitions that define your worst parts, but the truth is that most of us work long hard hours trying to hold on to and maintain the quality of what we already have. We desire for our lives *not* to be disrupted. When natural and man-made disasters strike or other mishaps undo our worlds, we crave "a return to normalcy," the reemergence of the everyday routines we struggle to sustain.

Since we began working on The Maintainers more than five years ago, people have consistently regaled us with (often disastrous) stories of domestic, or private, maintenance—what happens in their homes, how they care for their loved ones, themselves, and their possessions. Showering, doing laundry and dishes, fixing things that break or hiring someone to do it, vacuuming and cleaning, getting a haircut, wiping baby bottoms, clipping the toenails of aging parents—the work is endless.

That's life. As we saw in the last chapter, the vast majority of people work as maintainers, but *nearly all* of us are maintainers at home. Even the megarich bathe their own bodies, one prays. And let's face it: Even though some people take great pleasure in gardening, tinkering, and making home improvements, these maintenance-based tasks are often a pain in the ass.

In this chapter, we'll explore the work of upkeep, care, and repair that takes place within the home and family life, including looking after our own bodies. Maintenance is not restricted to public infrastructure and the priorities of corporations; it does not stop when we walk through the front door of "home sweet home." Many of the underlying factors that lead to maintenance problems in our communities and organizations, including the habit of deferring maintenance, also play out in domestic life. If we are to restore the role of maintenance in our society, we must understand the full sweep of the labor of upkeep and how it affects our private lives.

MAINTAINING OUR BODIES, MAINTAINING OURSELVES

When the journalist Stephen Dubner interviewed us for an episode of *Freakonomics Radio,* he said that bodily maintenance was the first thing he thought of after he'd read one of our articles. As you get older, he pointed out, "You spend more and more time maintaining yourself." Maintenance is the war against entropy—not only in technology but also in biology. Bodily maintenance is a constant part of human life, whether in the form of diet, exercise, or grooming. (Of course, many nonhuman animals clean and preen themselves, too.)

There is a wide divergence between the amount of work, time, and money that men and women put into self-maintenance. The reason, of course, is that we have wildly different standards for women's and men's appearances. In one poll, 81 percent of women reported using at least one beauty product in their morning routine while 54 percent of men reported using none of them.[1] This difference leads to widely divergent costs, too. Over a lifetime, men spend roughly $176,000 on beauty products and services, while women spend more than $225,000.[2] As one headline put it, THE AVERAGE COST OF BEAUTY MAINTENANCE COULD PUT YOU THROUGH HARVARD.

Beyond such double standards, some bodies simply require more care than others. Disability activists and scholars have pointed out that people with disabilities and the people who care for them do a tremendous amount of maintenance work just trying to maintain quality of life. Much of this bodily maintenance fits the definition of "dirty work" we encountered in the last chapter. It involves bodily fluids and parts of the anatomy that are rarely seen in public.

The disability studies scholar Hanna Herdegen has written about one example. A few years ago, a new trend took off on YouTube and social media. Lifestyle vloggers began adding the tag "What's in My Bag" to videos, and would pull items out of their backpacks or purses one by one while describing them. Vloggers with disabilities and chronic illnesses gave the tag a twist, creating a "What's in My Bag: Chronic Illness Edition." As they unpacked their bags, they would unveil all of the tools of self-maintenance they require that nondisabled people—or "ableds" as people with disabilities sometimes jokingly call them—would never dream of carrying around, like "feeding tubes, pulse oximeters, and blood pressure cuffs." Individuals with wheelchairs or prosthetics also often tote tools, especially screwdrivers, so they can perform repairs on the go. As Herdegen points out, even objects likely to be found in a nondisabled person's bag—sweaters, snacks, and yoga mats, for instance—perform different self-maintenance functions for the disabled and chronically ill.

Bodily maintenance and care may seem quite different from upkeep focused on technologies, but we've found that the same basic problems affect both forms, including discounting—or putting things off for the future in the name of upfront pleasures and gain—a growth mindset, and a failure to address widespread, collective problems.

It's no surprise that many people put off eating well and keeping fit for some imaginary tomorrow that never seems to come. We discount future costs and benefits in favor of those in the present, and this short-term impulsivity can undermine our long-term health. The costs are clear: More than 60 percent of Americans are medically overweight, and more than one-third are obese.

At the same time, it would be wrong to emphasize individual choices *too much* when it comes to the way we eat and the amount of physical activity we get. Earlier peoples got more exercise not because they were especially virtuous but because their livelihoods—principally the farming work that dominated human existence for most of history—*required* them to move. Furthermore, research suggests that when people experience scarcity through poverty and stress, they become more impulsive and make worse choices.[3] As we saw in the last chapter, nearly 40 percent of Americans struggle to make ends meet, and likely even more individuals are stressed out in other ways. Moralizing about self-maintenance in such a context would be cruel.

It's hard to encourage self-maintenance in a world that is literally designed to appeal to our worst selves. But what makes it even harder is that our ideal goals are not clear. Americans spend $33 billion a year on weight loss products, yet, as a rule, dieting does not work. Moreover, our culture obsesses over fat, leading to bullying and people torturing themselves psychologically, but weight is not as tied to good health as we have been led to think—from one-third to three-quarters of obese people are metabolically healthy, while as many as one-quarter of skinny people are unhealthy along the same lines.[4]

That is, there are more reasonable and grounded ways to think about self-maintenance than obsessing over weight, but this is not what you hear, in part because there are enormous industries—from magazines and books to dietary supplements to fitness clubs to ready-to-eat, portion-controlled, "lite" microwave dinners—that profit from your thinking unrealistically about maintaining yourself. This is the growth delusion we've seen in previous chapters in reverse.

The medical anthropologist Theresa MacPhail told us that she recently became aware of a similar growth delusion: the obsession with "gains" in the exercise advice world. MacPhail has been a runner for years, but when she turned forty-seven, she started thinking more broadly about her long-term health and decided to add some weight training to her regimen. She started the way any academic nerd would: by reading a ton about the topic.

But the more she read, something dawned on her: "After about the tenth article, I was suddenly like, '*Wait a minute,* this is total bullshit!' All the focus is on 'gains' and 'increases' in whatever you're measuring (mile time, endurance, weights, whatever). And, of course, I'm thinking to myself, 'It's a human body with decreasing capacity to repair itself. For the most part, you are just trying to maintain yourself. Why is maintenance being framed as a gain?' And then it hit me that fitness NEVER talks about maintenance. It's always focused on improvements.

"I guess, as I age, I'm really more agitated over our cultural refusal to admit to any limits, especially when it comes to our bodies. I'm pissed off that I'm expected to lift weights, do cardio, get some stretching in, do mental puzzles or learn a new language to keep my brain cells 'active' (whatever that means), make nutritious food, and get eight hours of solid sleep. If I did everything I was 'supposed' to do, I'd be exhausted and still getting older and declining. I guess what I want is a lot more admission that things break—including bodies—

and that's natural and normal. All we can hope for is that we slow the roll down that hill."

CARING FOR OTHERS

As Theresa MacPhail points out, caring for yourself can be a lot. But most of us also care for others. Popular culture is full of images of formerly hip, well-dressed parents looking gaunt and shattered, reduced to wearing sweatshirts and yoga pants, wiping baby butts, doing endless laundry, extruding Cheerios and banana gook from the deepest recesses of car seats, trying to find the will—just this once—to feed the kids something healthier than mac and cheese. Every parent tells these horror stories. Like when a little one vomits in bed at 2:00 A.M., and one parent attempts to hose off a screaming child in the shower while the other deals with bedding covered in carrot chunks, curdled milk, and the toe-curling smell of bile. Of course, all of this is much, much harder if you are a single parent.

In *Forced to Care,* the historian Evelyn Nakano Glenn points out that the burden of caregiving falls unequally on the shoulders of women, especially women of color. (The same is true for housework, as we'll see in a bit.) Seven out of ten informal (or unpaid) caregivers are women, and women in informal caregiving roles who also work full-time jobs do an average of sixteen hours of additional unpaid care work per week. Married women with children also do an average of fourteen hours of childcare compared to eight hours for dads.[5]

Intensive caregiving is stressful and is associated with a number of negative health outcomes, including heart disease, high blood pressure, diabetes, and depression. Such informal work also has significant financial tolls. One study found that individuals—again, mostly women—who made career adjustments to spend more time giving care to family members lost $659,139 potential earnings over a lifetime. Women who began caregiving early in life were two and a half times more likely to end up in poverty.[6]

Of course, some people either must hire other people to care for their loved ones or choose to do so to make life easier. But paid caregivers do not fare much better than unpaid ones. Women make up 90 percent of paid caregivers, and for a variety of historical and sociological reasons, many who do this work are women of color and/or immigrants. Home caregivers made $9.22 an hour in 2008, less than the federal poverty level, and they typically do not have benefits, vacation time, or health insurance.[7] They make so little they must take on multiple clients, working long, hard hours. The people who care for our loved ones often live stressful, precarious lives.

The philosopher Nancy Fraser argues that care work has changed a great deal over the last two hundred years.[8] The survival of our economy has always depended on unpaid care work that lies outside the job market. As Western societies industrialized throughout the nineteenth century, influential figures put forward an idealized image of domesticity that included a notion of "separate spheres" between men's paid work outside the house and women's unpaid housework. Few workers were paid enough to turn ideals of domesticity from fantasy to reality, however. As the welfare state emerged over the course of the twentieth century, advocates pushed for a "family wage" that would enable the breadwinner's income to support all of the unpaid work the family required. Many families lived far away from this ideal during this period, but the standard of living rose for most Americans and for people living in other developed countries. Since the 1980s, however, politicians have cut back social welfare, and wages have stagnated. Meanwhile, women have increasingly moved into the workforce. Fraser argues that these shifts in the social safety net and employment have led to today's "crisis of care." Many families struggle with the questions of who is supposed to do the care work and how it is supposed to be paid for. And for many, the only available answer is to keep their shoulders pressed snugly to the grindstone.

KEEPING UP A HOME

Ah, housework. If your life is anything like that of many of the people we talk to, your week goes like this: You work from Monday through Friday and come home in the evenings feeling too exhausted to do much of anything. Over the course of the workweek, your home exponentially becomes a disaster zone. If you have children—otherwise known as Agents of Chaos—the exponents have exponents; by Friday night, when you pour yourself a big glass of something strong, you are ankle deep in clothing, toys, and the little ones' precious works of "art." You will yourself toward sanity as you attempt—but mostly fail—to maintain some semblance of order.

The amount of housework that men do has gone up over the years, but women still do more. And while same-sex couples tend to divide household chores more equitably, things become less equal and more like heterosexual couples if they adopt or have children.[9]

How maintenance works in any given household depends heavily on how much money it has. Wealthy people hire help to keep up their homes and yards. Drive through any U.S. city and you can tell where the rich live and where the poor live first and foremost by how the houses and landscaping are maintained: that golf-course-perfect American lawn of the well-to-do versus the crabgrass-infested, bare-spot-pockmarked yards of the poverty-stricken. In some neighborhoods, homeowners' associations pressure households to keep entropy in check, lest newcomers let standards slide.

Most families cannot afford help. They make do on their own. But as we've seen in other contexts, putting off or deferring maintenance work is a constant temptation. Both of us have become first-time homeowners in the last few years, and have come face-to-face with the crushing maintenance work owning a home can entail. When Lee bought a home in Blacksburg, Virginia, he and his wife hired a local, well-known, and highly regarded home inspector named Bob Peek. Peek has examined thousands of houses around the New River Valley region. It's a calling, he told us.

Peek became a home inspector after reading a news story about a local man who'd purchased a home only to realize that something was disastrously wrong with the house's foundation. The problem was hidden behind a board in the basement. If the man had gotten a home inspection, he never would have had an issue, but as it stood, the problem became a calamity. The man became financially insolvent. He got divorced. His life was ruined. Peek realized that he could help people like that out and that doing so would probably be rewarding.

When we asked Peek if we could interview him for a book about maintenance, he replied, "Well, I'll tell you one thing. People don't do it! They'd rather go to Disney World than put money into doing preventive maintenance on their home." As a home inspector, Peek sees the absolute worst of deferred maintenance. Homeowners fail to do even the simplest things, like change out HVAC filters and repair their visibly degraded roofs. Recently, Peek put his foot through a roof while doing a home inspection. An elderly woman hadn't maintained the roof, and a leak had rotted out the wood. Luckily, Peek was not badly hurt. "It's the decks that really scare me, though," he told us, describing entire structures wobbling as he walks across and inspects them. "I have nightmares of them collapsing during some family gathering."

Deferring home maintenance is a problem, but it's been exacerbated by how the size of American homes has grown since World War II. In 1950, the average house built in the United States was 983 square feet; by 2014 it was 2,657 square feet.[10] As our houses have grown, so has our debt load. U.S. household debt hit $13.2 trillion in 2019, or 21 percent above post–2009 financial crisis levels.[11] (A big part of the increased debt came from car loans; mortgage debt, which sat at $9.1 billion, was about equal to 2009 levels.) A bigger house and more stuff also means more maintenance, though we often underestimate these long-term costs when making the initial purchase.

One of our colleagues recently moved to a semirural college town after living in a major U.S. city. She was shocked by how much cheaper the home prices were in her new town, so she bought a big home with several acres, enough to keep horses and some other farm animals. A few years later, however, she got divorced and suddenly had to maintain or pay others to maintain the house, the barn, the horse fences, and all the land. The work overwhelmed her, but in a way she was stuck: There weren't any homes for sale in the area that met her needs. The best option was to keep pushing on. She was lucky: She had the means to do so . . . barely.

Problems with home maintenance are widespread and have very real consequences, *especially* for the poor. Melissa Jones, an expert at the Virginia Center for Housing Research, explained to us that deferred maintenance is a significant problem throughout the United States, including urban, suburban, and rural spaces. Baby boomers have experienced stagnant wages since the 1980s, and as a result they have not kept up their homes. When millennials and other young people buy property today, they often walk into structures that need $40,000 to $50,000 of rehabilitation and maintenance, costs that are not necessarily visible up front, even with home inspection.

Things are even harder for families that live in poverty or struggle financially. Some people are "house poor"—they own their homes, but most of their income goes toward their mortgage, leaving them too impoverished to care for the structures. Work done by the Habitat for Humanity of the New River Valley in Christiansburg, Virginia, has uncovered some of the extent of the problems in nearby regions.

Habitat for Humanity is known for building houses for low-income families. But when the financial crisis hit in 2008, it challenged the organization in two ways: First, people stopped giving donations; second, the families living in Habitat's houses stopped making the mortgage payments on which further homebuilding work depended.

Increasingly, the Christiansburg Habitat for Humanity came to focus on doing repairs for needy residents and helping individuals do their own repairs and upgrades. Through this work, they started to get a sense of how great the need for maintenance actually was in their community. As we will see in chapter 11, Habitat has started holding so-called repair cafés and creating a tool library that allows individuals to borrow tools they don't own.

In 2015, the Christiansburg Habitat for Humanity worked with its partners to raise a little more than $83,000 to provide "aging in place" home modifications for local residents in the New River Valley, a region with a population of ninety-five thousand. When they put out a call for applications, the selection criteria were fairly restrictive: Applicants needed to be over fifty-five and eligible for low-income assistance, with preference given to women living alone, individuals in poor health, and wheelchair users. The program hoped to complete twenty-four home modifications over two years, an average of one per month.

As a report later put it, resident response was "unexpected and overwhelming." The partners received 106 applications in the first six months. One noteworthy feature was that many applicants did not fit the criteria, often because they were younger than fifty-five, yet these people were experiencing real needs in their homes and so applied anyway. As members of the program worked through the applications they realized something else: While the program was meant to improve homes to enable the elderly to age in place, many of the dwellings had critical repair problems from years of deferred maintenance that went well beyond the scope of the program.

Carol Davis, the sustainability manager for the town of Blacksburg, told us about a few recurring causes of home maintenance problems: Men who had taken care of home maintenance in previous decades stopped as they grew older and their health declined; men were sometimes too proud to admit their capacities had slipped; after the death of a man responsible for maintenance, his widow or former partner either wasn't aware of work that had been done or did not have the skills to continue the work.

But the biggest problem was money. Program leaders found that many applicants were in the process of, as the report put it, "getting poor." They were "falling into poverty after a lifetime of living in the middle class. Their descent was typically caused by the death of a spouse and a loss of income, or they experienced a significant and costly healthy problem," which left them in a "crippling" financial situation. "I've met elderly women who were literally praying that they'd die before their home crashed in around them," Shelley Fortier, the executive director of Habitat for Humanity of the New River Valley, told us.

Younger applicants were in the same situation because they started poor. One had plans to work and earn an associate's degree at a local community college, but these plans fell apart when she gave birth to a daughter with intellectual disabilities. Initially, the woman assumed that insurance and the public school system would cover her daughter's care and education, but she soon found herself spending more each month than she made at her part-time job. As the report describes, "She quit her job so she could be with her daughter, stopped attending college, and moved into a cheap home that she thought she could fix up. Now she realizes that she can hardly make ends meet each month. Her home needs a lot of work and she fears she could find herself homeless if anything more goes wrong." There was no money to fix the house; small problems, like a tiny hole in one portion of the roof, could spiral out of control and make the home unlivable.

Others were living on the edge, too, with broken or run-down home systems, like heating, air-conditioning, even water. The program uncovered particularly bad and exploitative conditions in trailer parks. As the report noted, "Mobile home park owners typically sell mobile homes situated on their land on a rent to own basis with high interest rates (upwards to 25%). The purchaser also pays monthly lot rent and water fees to the landowner. Once a tenant fails to make a payment on the trailer, the trailer reverts back to the owner, the occupant is forced to leave and loses all the equity they put into the home. Trailer park owners resell the mobile home to another individual without making any upgrades or improvements."

There's every reason to believe that many people who could have used these home modifications and repairs did not apply because they realized they didn't meet the age and income criteria. But even if you take the 106 people who did apply for help in Virginia's New River Valley as somewhat representative of residents who have critical home repair needs, it suggests many hundreds of thousands, perhaps millions, of households across the United States are in this position.

Renters and people living in public housing face other kinds of maintenance problems; often they have no control over their situations. In Washington, D.C.'s Columbia Heights neighborhood, one building's residents went on a rent strike after their landlord refused to improve their hideous living conditions that included rats, cockroaches, and other vermin; leaking walls and black mold; failing heating systems; and electrical wiring that often shorted and sparked.[12] It's likely that the landlord was trying to use the conditions to encourage the residents to flee rent-controlled apartments, so that he could sell the building for more money. But many other tenants face more mundane headaches when landlords either delay or refuse to do repairs or do them shoddily.

Things can be even worse for public housing residents. The sociologist Daniel Breslau has found that pest control in public housing not only doesn't kill cockroaches and other disease-carrying pests once and for all—it isn't even meant to do so. The chemicals sprayed briefly push the critters back but also enable their return, guaranteeing perpetual business for the pest control companies. Public housing residents face all kinds of other maintenance problems. Perpetually broken elevators are perhaps the most famous example, creating outcry and controversy in New York and other American cities. When the elevators are always out of service, the elderly, disabled, and other residents with mobility problems become near-shut-ins, relying on others to pick up groceries and other provisions for them.

Welfare recipients, renters, and homeowners face maintenance problems for different reasons and in different ways, but the common denominator for all is the life-disrupting consequences of such problems. What Shelley Fortier and others are doing to try to change this situation is discussed in chapter 11. Next we will examine how the maintenance problems that affect homeowners apply to the consumer products we use every day. Corporations have purposely made it even more difficult to repair appliances and electronic devices than it is to fix our homes.

CONSUMER GOODS AND THE RIGHT TO REPAIR

One of the most remarkable things about our society is how relatively few of our broken possessions most of us will repair.

We have become a culture of disposability. It is hard to overemphasize how new this is. For most of human history, objects were both produced and maintained locally. Blacksmiths, for instance, not only forged items like knives, axes, and axles for horse carts but also kept them in decent condition. Fashions changed slowly, and people made their clothing and furniture to be durable, often repairing and preserving them for a lifetime or more. As the historian Rosalind Williams notes, "In some parts of the world, the basic garment of the common man went unchanged for centuries—the poncho in Peru, the dhoti in India, the long shirt in China, the kimono in Japan."[13] In such a context, "possessions were handed down from generation to generation."

Mass production changed all of this, first and foremost by driving down prices. At first, decreasing prices meant that even average people could afford consumer goods like electric toasters, radios, and televisions. But the process continued, eventually making objects so cheap that we came to view them as disposable. It's now cheaper to buy a new toaster than to repair one that is acting up.

The easy availability of inexpensive goods has shaped our everyday experience. The fact that we can go to a big-box store or press "buy now" on the website of an online vendor and fill our homes with inexpensive objects is near magic. Average Americans today own quantities and qualities of goods that only the wealthiest individuals could afford a century ago. Even relatively poor people own mounds of cheap crap that mostly goes unused.

Over the course of the twentieth century, new homes were built with ever-increasing amounts of storage room, including the advent of the walk-in closet. While garages were originally added to properties with the purpose of storing cars, 93 percent of Americans now use their garages as storage spaces, and 30 percent of homeowners do not have enough room in their garages for a car.[14] Moreover, in recent years, the self-storage industry has grown in leaps and bounds, becoming a $38 billion a year market that one in eleven Americans uses.[15]

Some people find this plethora of goods overwhelming. The organizing guru Marie Kondo's book *The Life-Changing Magic of Tidying Up* became a mega-bestseller when it was published in the United States in 2014. Some have argued that Kondo's decluttering philosophy only works for the privileged; it's easy to throw things out or give them away when you are confident you can replace them whenever you want.[16] But the reception of Kondo's bestseller showed that many Americans were becoming frustrated with the piles of cheap stuff that overflowed from their closets, garages, and lives.

In the meantime, even when an object in our life *is* expensive enough to repair, it has become much more difficult to do so. Much of this difficulty is because computers have been built into so many things around us—most notably our cars. Automakers first put computers in cars in the 1980s and '90s to meet federal air pollution standards, but the companies soon saw strategic potential in the technology: They could use computers to monopolize repair and force owners to go to dealerships to get work done. Consumer advocates call these corporate strategies "repair restrictions."

By the early years of the twenty-first century, aftermarket companies, including local auto mechanics and parts stores, were seeing their business drop because of the restrictions. They began to lobby Congress for an automotive right-to-repair law, but their efforts hit a dead end. After turning to state legislatures instead, they eventually found success in Massachusetts, which passed the first automotive right-to-repair law in 2012. The bill required car companies to make the same vehicle diagnostic and repair information available to independent shops that they gave to their dealers and other authorized facilities. The automakers caved. Fearing the multiplication of laws across different states, they agreed to make the Massachusetts law an industry standard.

By 2012, repair restrictions had moved well beyond automobiles, however, as other manufacturing sectors saw new business potential in controlling fixes.

At about that time, the so-called right-to-repair movement emerged to fight repair constraints. Kyle Wiens, the CEO and editor in chief of the online repair guide website iFixit, is one of the leading voices of the movement. Wiens got into repair when he dropped and broke an Apple iBook G3 when he was an undergraduate at California Polytechnic State University in 2003. He decided to fix the computer himself, but when he couldn't find a repair manual online, he posted a how-to guide on his webpage. Surprised by how many views the videos got, Wiens and his friend Luke Soules started iFixit, a company with the goal to "teach everybody how to fix everything."

Wiens later learned that Apple was using the Digital Millennium Copyright Act to force people who posted its repair manuals to take them down. Apple controlled how consumers could fix their devices in other ways, too. For years the company claimed that if consumers had their iPhones repaired by a local repairperson, it would void the warranty, because it might harm the phone and prevent it from being repaired properly in the future. Right-to-repair advocates argue that such warranty rules violate the Magnuson-Moss Warranty Act of 1975, a federal law meant to protect consumers from unfair or misleading warranty practices.

It's difficult to measure how widespread repair restrictions are, but the U.S. Public Interest Research Group (U.S. PIRG), a consumer advocacy nonprofit, has started working on the topic in recent years. Nathan Proctor, who leads the organization's right-to-repair efforts, surveyed fifty companies that are members of the Association of Home Appliance Manufacturers. He found that forty-five—or 90 percent—of them claimed that third-party repair would void their warranties, a stance that may violate federal law.

Small businesses, individual consumers, and environmental sustainability are all negatively impacted by repair restrictions. Nathan Proctor recounts the story of a man who operated a tour boat business in the southern United States. The man owned a large boat that was capable of carrying more than forty passengers and was powered by a Volvo diesel engine. When he put the boat in the water for the season, he found that the engine would not go any faster than an idle. It turned out that a Volvo repairperson needed to "certify" the repair that had been done locally in the off-season before the boat would be capable of higher speeds. But the closest Volvo service technician was busy and lived four hours away, so it took more than a month to get the work done. Half the boating season was over, and the tour boat owner was out tens of thousands of dollars. Only extreme economic resourcefulness kept him in business.

Repair restrictions raise costs for consumers, too. For instance, because Apple has barred you from taking your device to a local potential competitor, it can charge you $1,000 more for a repair.[17] As we saw in chapter 6, Stephanie Hoopes of the United Way estimates that about 40 percent of American households struggle to make ends meet. An unexpected expensive repair is precisely the kind of thing that can push one of these families over the edge.

Right-to-repair advocates also highlight environmental sustainability and the value of community as important causes. Many electronic devices contain rare earth minerals and other nonrenewable resources and yet companies design them to be disposable, unrepairable, and unrecyclable. For instance, Apple has long made unrecyclable products by gluing glass to aluminum, turning both materials into waste. A recent piece on Vice Motherboard called Apple's AirPod headphones a "tragedy": Both unrepairable and unrecyclable, they cannot be thrown away because their lithium-ion batteries are known to cause fires.[18]

Moreover, electronics firms only support products for so long. In the early twentieth century, companies like General Motors introduced "planned obsolescence," using techniques like annual model changes to encourage consumers to keep consuming. But some have recently argued that electronics companies have pioneered something we might call "forced obsolescence."[19] After so many product cycles, companies simply stop supporting and updating their products. Unless a user has the skills necessary to keep the systems running, a perfectly operational product will die. Nathan Proctor of U.S. PIRG points out that Americans throw out 416,000 cellphones a day.

We have become a rich culture—unprecedentedly so in all of human history—in large part because of the material realities that make up our daily lives. (Though, for sure, that wealth has always been unequally distributed.) At the same time, we have become a culture that is profoundly disposable, unmaintainable, and unsustainable.

In the last couple of chapters, we've seen how several factors, including our culture's obsession with innovation-speak and unwillingness to face reality, negatively affect maintenance in several domains of human life, from large-scale public infrastructure to public and private organizations, from the daily labors of maintainers to our everyday lives with technology at home. Frankly, it's a grim picture, and many people have told us that they look upon it with despair. We understand that pessimism and always strive to be realistic. But we also believe there are reasons for hope, a hope reflected in the many incredible people we've encountered in the last few years of studying this crisis.

We know it is possible to improve how we maintain public infrastructure and technologies within organizations, to better recognize and compensate maintainers, and to live a saner and more humane life of maintenance and care within communities and at home. Smarter approaches to management, better ideas in public policy, and more realistic approaches to the ways that individuals work and live their private lives will be required. Bringing people together and mobilizing collective action to infuse the world with richer spirits of maintenance, care, and sustainability are crucial to success. Part 3 shares stories about the people whose ideas make us optimistic.

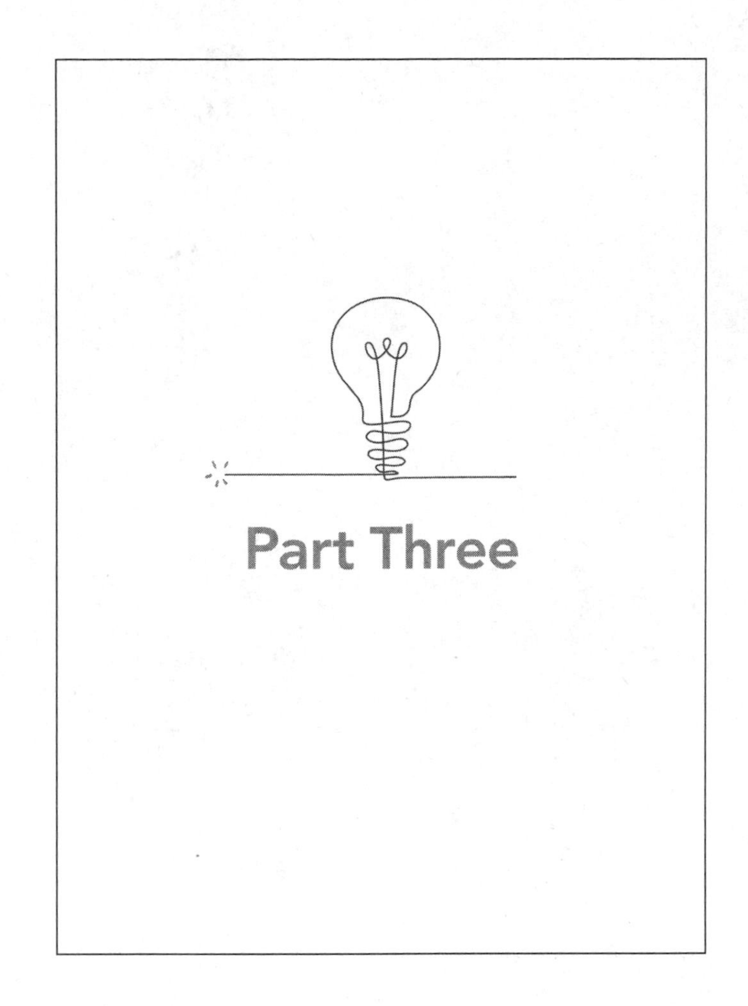

Part Three

CHAPTER EIGHT

The Maintenance Mindset

RESTORING A CULTURE OF CARE

To this point, our book has explored the devastating consequences of going all in on innovation. One of those consequences is a failure to maintain the things that matter: our technologies, our bodies, our communities, and the fundamentals that allow our businesses to succeed.

That's the bad news. The good news, which we'll spend the rest of the book sharing, is that there is another path forward. It starts with an awareness of the importance of maintenance, a commitment to keeping things in good working order, and attention to the investment of time, energy, and resources required.

In this chapter, we'll meet several people who have put this mindset into action and championed the cause of maintenance. Considering the bad actors profiled elsewhere in this book, it might be tempting to call them "heroes." But in many ways, they are antiheroes: They reject the idea that maintenance can be achieved through heroic acts, prefer to work behind the scenes, and measure success as what happens when things keep functioning as intended. They will tell us that we don't need heroes of maintenance, so long as we utilize good planning, hard work, and, every now and then, a little bit of ingenuity.

To understand the maintenance mindset, you should start with a question: What is good and worth preserving? This is the fundamental point of departure from the language of innovation, which asks you to worry about what you need to change, or what will be disrupted—it is a language of fear. Instead, we're asking you to adopt a new habit when you walk around, or think about your work, your community, and your personal life: Ask yourself, What is good here? And how can I maintain that goodness? How can I preserve and extend that which is valuable?

These are the questions that we've focused on in our research, in our interviews with successful maintainers, and at the conferences where we've brought together people who are passionate about upkeep and care. From all of these conversations, we have distilled three general principles of the maintenance mindset.

First, there is the principle that **maintenance sustains success.** Maintenance consists of activities that, when done correctly, ensure longevity and sustainability for a company, a city, or a family home. To put the point a different way, no innovation can persist without maintenance. Second, there is the principle that **maintenance depends on culture and management.** Good maintenance is possible only with good planning that takes an organization's preexisting culture and values into account. The third principle is that **maintenance requires constant care.** The best maintainers take a nurturing and supportive approach to their work. They are often detail oriented, creative, and, more than anything else, dedicated to their craft.

We'll explore each of these principles in turn, and show you how maintainers apply them to keep businesses and society in good working order.

MAINTENANCE SUSTAINS SUCCESS

It may be obvious that any serious attempt at maintenance requires investment. What is less obvious is how much return on investment executives or managers should expect. Anyone with limited resources may be sympathetic to the moral arguments in favor of maintenance—sure, we should maintain and preserve everything that's important to us. But most of us can't make decisions driven purely by moral arguments. We also need to get real about costs.

Because they are sensitive to the importance of these questions, maintenance and reliability professionals spend a lot of time documenting return on investment. One such study was published by Jones Lang LaSalle, a commercial real estate services firm that broke into the Fortune 500 in 2015 and has since risen to the 189th position with more than $16 billion in annual revenues. The study analyzed costs at a "large telecommunications firm," measuring expenditures, the frequency of repairs, the effects of maintenance on energy consumption, and the various types of equipment across nearly fourteen million square feet of the company's facilities. Fifteen categories of equipment were analyzed, from air compressors and handlers to roofs and parking lots, and the Jones Lang LaSalle consultants factored in costs of repair, replacement, downtime, and energy consumption.

The study reached some astonishing conclusions: Preventive maintenance programs had generated a whopping 545 percent return on investment. Most of the value came from how proper maintenance extended the lifetime of the company's equipment. To take a single piece of equipment—an air chiller—as an example: The average replacement cost for a chiller was roughly $350,000; an annual maintenance contract was $5,500. Extending the useful life of chillers not only delayed the costs of replacement but also avoided the costs of emergency fixes or replacements and contributed to energy efficiency, a benefit that would show up elsewhere on the company's books.[1]

The Jones Lang LaSalle study was a landmark effort to quantify the value of preventive maintenance, but it falls into a familiar genre of success stories that maintenance professionals like to share. Two examples come from Augury, a predictive maintenance firm that specializes in sensors and data analysis. The first example comes from a client who operates a facility that manufactures large home appliances, where Augury's sensors identified a severe malfunction in a compressor that other vibration analysis equipment had missed. Because the fault was identified before the compressor failed, the company spent only $7,000 repairing it. If the compressor had failed, the replacement would have cost an estimated $240,000 in parts and downtime. The second example comes from another client, a medical device manufacturer, where Augury's equipment identified a malfunction in an air-handling unit and pinpointed the failure to a motor that cost $3,500 to repair. The facility manager estimated the overall cost of failure would have exceeded $200,000 in repair and downtime costs had the fault not been identified prior to catastrophic breakdown.[2]

These cases illustrate the enduring relevance of the aphorism "An ounce of prevention is worth a pound of cure." Nevertheless, it can be a struggle to convince executives that they should budget for preventive maintenance. Andrea Goulet, CEO of the "software mending" firm Corgibytes whom we met in chapter 1, has found that homeownership is the most effective metaphor to get this point across. She told us, "It's cheaper to repair a roof when you first notice it leaking than to wait for it to come crashing down on you."[3]

Maintenance professionals—mechanics, engineers, nurses, janitors, building superintendents, and so on—have shared many dozens, even hundreds of these types of stories, in which maintenance generated significant returns on investment. The common threads among them are that the single most important enablers of these returns are *attention* and *commitment*. In other words, there's no magic piece of software or hotshot consulting firm that can swoop in to save the day. Organizations that turn to software or consultants to realize savings from proper maintenance will fail unless they make underlying changes in culture and administration. It is these changes and attitudes that we refer to collectively as the maintenance mindset.

Maintenance is such a powerful tool in sustaining success because almost any definition of success assumes that the success be enduring. What's the point if your accomplishment lasts only a moment? Some organizations want to achieve profitability; others want to enhance safety and reliability; others want to protect a cultural heritage of some kind; still others wish to advance their particular vision of justice; and so on. For all of these goals, hard-earned triumphs will be better if they are sustained over an extended period of time—in other words, if they are maintained.

People value maintenance because it ensures reliability. The reliability of software, for example, has been a major concern for computer professionals ever since software was first developed. The cloud is useful only if it's reliable, and users of network services are unhappy when those services are down (at least this is what we experience when Hulu drops in the middle of a movie, and our kids have meltdowns). We use chat and email to keep in touch, video and photo apps to savor moments and pass the time, and calendars and checklists to keep us on schedule. And although crashes and bugs are the stuff of everyday experience, it's worth pointing out that some elements of our digitized existence are so reliable, and so well maintained, that we hardly notice the effort that goes into their persistence.

The corporate icons of our digital civilization—Google, Amazon, Facebook, Apple, and so on—all have deep investments in maintenance and reliability. They compete on uptime, and in this respect, successful digital firms are heirs to their predecessors in the railroad, oil and gas, and telephone industries. These companies became successful when they embedded maintenance and reliability into their organizational routines. As we saw in chapter 3, when we looked back to the Roadmasters' and Maintenance-of-Way Association of the late 1800s, professionals in those industries devoted substantial time, resources, and ingenuity to ensuring reliability.

A twenty-first-century counterpart to the Roadmasters can be found within the community of developers at Netflix, who developed an approach to network reliability testing that they call "chaos engineering." Here's how it works: Netflix engineers built a tool, called Chaos Monkey, which operates within Netflix's production network. The tool simulates random disruptions to test how the overall system responds and encourage developers to design new features with a high degree of resilience.

Yury Izrailevsky and Ariel Tseitlin, formerly Netflix's directors of Cloud Solutions and Systems Architecture, compared their challenge to the experience of a driver who gets a flat tire: "Even if you have a spare tire in your trunk, do you know if it is inflated? Do you have the tools to change it? And, most importantly, do you remember how to do it right?" The healthiest response, in software as in road safety, is to be proactive. "One way to make sure you can deal with a flat tire on the freeway, in the rain, in the middle of the night," Izrailevsky and Tseitlin continued, "is to poke a hole in your tire once a week in your driveway on a Sunday afternoon and go through the drill of replacing it."[4]

If you feel some comfort knowing that teams of engineers at Netflix are doing everything they can to ensure there's no jitter when you're watching *The Epic Tales of Captain Underpants,* then you might feel more comfort knowing that the people entrusted with the physical safety of the entire nation also take maintenance and reliability very seriously. In 2015, the Department of Defense designated 40 percent of its $500 billion budget for operations and maintenance.[5] "Operations and maintenance" is the largest single category of appropriations within the DoD, more than weapons purchases, employee compensation, and the construction of bases, facilities, and housing. In addition to the DoD's properties, there are also the costs required to sustain weapons systems. A 2018 *New Yorker* article provided a glimpse: "In 2016, the maintenance contract for the Royal Saudi Air Force's two hundred and thirty F-15 fighter jets alone was worth $2.5 billion."[6]

As you might imagine, accounting for expenditures at this level is a significant endeavor unto itself. Congress, as well as various branches within the DoD, complete a number of reports on the various pieces of the puzzle, such as a Congressional Budget Office study of depot-level maintenance of F-35 fighter jets. Another CBO study from 2017 estimated that operation and maintenance costs for the Department of Defense grew by almost 50 percent between 2000 and 2012, driven by increasing expenditures on healthcare, civilian pay, and contract services for maintenance and operations. In an era when Congress has imposed significant spending constraints on military budgets, these increases have raised eyebrows on Capitol Hill and at the Pentagon.[7]

For the military, the consequences of poor maintenance can have a serious impact on its core mission. Indeed, the question of what constitutes "success" can be a tricky question to answer for an organization like the Department of Defense, with annual expenditures well into the hundreds of billions of dollars. However, at a certain level, the answer is obvious: Success means military readiness, the ability to provide for the national defense.

So it's understandable that alarm bells have been ringing at the Pentagon in light of reports that the military's material and human infrastructure is deteriorating. The Pentagon's chief infrastructure officer told Congress in 2018 that the DoD has more than $100 million in deferred costs for facilities maintenance and restoration, with 23 percent of its facilities rated "poor" and another 9 percent rated "failing." The costs of the past are catching up with the U.S. defense establishment, in the shape of facilities that have reached the end of their useful life; remediation and retroactive cleanup costs for contaminated groundwater at military bases; global changes in military strategy that have rendered obsolete the massive Cold War–era "superbases" that are scattered around the globe; and the ever-increasing costs of dealing with climate change, which manifest in severe weather events such as hurricane damage at Tyndall Air Force Base in Florida, and flooding damage at Offutt Air Force Base in Nebraska. As of 2019, the best-case scenario for planners was asking Congress to approve a funding increase so that the Pentagon could make strides in dealing with its infrastructure and deferred maintenance backlog over the next *three decades*.

There is a paradox at work here: How can the organization with the largest maintenance budget on the planet still generate the feeling that it's not spending enough on maintenance? To get back to the question we posed earlier: What constitutes "success" for the U.S. Department of Defense? More than simply providing for the defense of the nation, or ensuring that facilities and matériel are in good working order, its success also includes the satisfaction of its employees and contractors, as well as the citizenry and elected representatives who allocate several hundred billion dollars in tax revenues to it. All of these groups are justified in wondering if the Pentagon's commitment to a maintenance mindset is as thorough and as genuine as it should be, given the resources and responsibilities it is entrusted with.

MAINTENANCE DEPENDS ON CULTURE AND MANAGEMENT

One really effective way to understand the maintenance mindset is to see the world from the vantage point of people who are immersed in maintenance—that is, the people who literally make maintenance their business. Like all professionals, maintenance experts attend conventions where they talk about possible problems, think about common solutions, listen to inspiring keynote addresses, and network with vendors and their peers. We had a delightful visit at one of these gatherings, the 2017 Mainstream Conference in Nashville.

At Mainstream, we were struck by the diligence and sincerity of everyone we met, as well as their passion for maintenance and reliability. We were likewise taken aback by the ubiquity of maintenance professionals, who work in all sectors of the economy and all facets of our society. For any industry or organization you can imagine, there are teams of professionals who handle maintenance and manage the facilities. From restaurants to sewage plants, from schools to prisons, from fracking to wind turbines, all organizations have assets that need to be maintained.

We'll return to this crucial word—"assets"—below. But before we get into those details, there was one phrase that stood out to us from the three days of the Mainstream Conference. It stood out because speaker after speaker returned to it, meme-like, in their varied presentations.

"The soft stuff is the hard stuff."

What did this mean? In short, it was an admission—a shared acknowledgment, really—that maintainers' most severe challenges didn't come from the technical and mechanical aspects of the job, such as servicing hardware or software interfaces, or the mysteries of random machine failures. Rather, the consensus was that the most profound difficulties in their work fell into the realms of what engineers like to call "soft skills"—communication, time management, acting as a team player, and so on. And this is where maintenance managers find their greatest challenges: convincing employees to change their routines, use new software, or be more courteous in their interactions with staff in other departments. There is training that can make maintainers more effective, but the most important element of a healthy maintenance culture is, at root, a matter of getting people organized and focused so that they can apply known techniques and strategies to tackle known problems.

At the core of professional maintenance is the concept of "asset management," the approach by which an organization takes care of an asset, or an "item, thing or entity that has potential or actual value."[8] At a high level, there are useful distinctions between financial, physical, and organizational assets—but they all share the quality of being essential for the efficient and effective performance of any organization. The notion of asset management, therefore, provides a holistic way for a company, or any organization, to think about and coordinate the things that it owns.

Asset management is important enough to be the focus of professional societies, conferences, and even international standards. The field has a number of widely respected consultants—experienced professionals who have implemented maintenance programs at a variety of facilities, and have wounds and scars to show for it.

Ricky Smith is one of these consultants. Like others experienced in maintenance and reliability, Smith takes pains to emphasize that it's not difficult to adjust a chain drive or grease bearings. The difficult work comes in changing human behavior. We're a stubborn lot, even when better approaches are staring us in the face.

A native of South Carolina, Smith first encountered maintenance while working as a mechanic in the U.S. Army during the Vietnam War. After he left the army, he became a maintenance technician for Exxon and then did the same job for Alumax, an aluminum smelting company in South Carolina. You may recall Alumax from chapter 3, where we described its pioneering role in "World Class maintenance" in the 1970s and early 1980s. At Alumax, Ricky Smith worked under John Day, immersing himself in Day's proactive approach to maintenance (including the 6:1 rule of planned to unplanned actions), the use of computer databases to manage tasks, and the approach to maintenance as a contributor to profitability that can generate measurable returns on investment. Through his work with the Kendall Company and later as a consultant, Smith estimates that he's advised well over five hundred companies around the world. He sees his expertise in maintenance as his way to make a positive contribution to the world: "All I want is for people to live a happy life and have equipment that runs the way it's supposed to run, and they're satisfied with it." His words carry the sincerity of someone who has spent decades helping people who are under a lot of pressure. "Life is a lot less stressful without equipment breaking down. That's very stressful, for everybody."

Smith tells the story of being asked by a friend, the president of a manufacturing company, to visit a plant that was losing millions of dollars a year. When Smith arrived, the local plant manager refused to meet with him. He laughed as he told us the story: "I called the president and I said, 'This place is totally screwed up. I mean one hundred percent. Production, maintenance, everything. People get stressed out working in a place like this. I'm firing you as a customer. Don't call me back.' And I hung up."

But the story didn't end there. "About six months later he called me back. And I said, 'Okay, I'll tell you what. You bring all your plant managers and maintenance managers down to Charleston, South Carolina. There's a park about a mile from my house. Your people will all stay at the park, in the rustic cabins. That way I can go back home if I don't like what I hear and see.' "

When the managers arrived a few weeks later, Smith read them the riot act. "I'm not here for the money. I'm here to help you guys. It's not about you. You're letting down the people that work for you. That's the ones I worry about. Your stockholders lose money because of you."

A skills evaluation was in order. "So the first thing I did was to sit them down and make them take a certified plant maintenance manager exam. Open book. That opened their eyes, and I said, 'Okay. Red alert.' In the end we had a great three days, and I helped them a lot. The president called me years later and said, 'Ricky, I didn't tell you this before, but you saved my job. You saved all their jobs. You saved the company.' " When we asked Ricky to reflect on the moral of his story, he said it was simple: "It's amazing how many people are out there suffering because their leadership doesn't listen."

There are thousands of people like Ricky Smith who work in the professional fields of maintenance and reliability. They are passionate about their work because they see how maintenance supports a great variety of things that people *really* care about—values such as safety, community, or justice. In other words, maintenance is an effective means to achieve greater ends.

For example, politicians in the United States are rallying around the concept of a Green New Deal, an equally ambitious and sweeping vision of economic transformation that, according to the Sierra Club, will "tackle the twin crises of inequality and climate change." Elsewhere, for-profit companies have adopted the ethos of the "triple bottom line" that measures progress in terms of social equity and sustainable environmental practices, alongside the more familiar "bottom line" of financial profit.

One company that's an especially useful example for our purposes is Fiix, a Canadian software maker. Fiix is in the maintenance business: It makes software for computerized maintenance management systems. We'll talk more about these systems later—what they are, how people use them, and what impact they can have on an organization. But for our present purposes, let's focus on Fiix's values, particularly the connections it makes between maintenance and sustainability.

Fiix's website makes this connection clear: "Maintenance and sustainability go hand in hand. Properly maintaining assets and infrastructure can help cut down on waste and emissions, and protect investments in infrastructure." Fiix also emphasizes that the company itself benefits from a focus on sustainability, because its "deeper sense of meaning and purpose for employees" helps it to "attract and retain top talent in technology."[9]

To understand how Fiix arrived at this point, we spoke to one of its founders, Marc Castel. His upbringing is certainly not typical for a software entrepreneur. He told us, "I grew up on a farm. For us, fun was taking apart the old tractor engine, building dune buggies, and then seeing if you could make new things out of those bits and pieces of stuff you had lying around."

He described his staff at Fiix as "jacked up by maintenance." In some ways, he built the company in the image of his boyhood self. "We're here because we feel like we have a corporate higher purpose, and that's to make the world more sustainable by enabling better maintenance and care of our stuff," he emphasized. "One hundred percent of my staff joined our company to be part of the sustainability revolution and to meet the crises of climate change, environmental degradation, resource consumption, and sustainability of the planet. They're just stoked about it, and they think maintainers are cool."

Marc's enthusiasm is infectious, and it was easy to see how users, investors, and employees could be persuaded to follow such a vision. "We're not a software company. We're a sustainability company," he explained. "If you think about sustainability and maintenance—in our minds they're the same word."

One employee at Fiix who translates this vision into numbers on the triple bottom line is Katie Allen, Fiix's manager of corporate social responsibility. Katie tells us that Fiix's culture of valuing sustainability played a role in the decision to join Fiix for 85 percent of its new employees. Its values also pay off in sales revenue: Fiix attributes 6 percent of its annual revenue to its corporate responsibility programs, and 20 percent of its customers in 2017 and 2018 combined told Fiix that these programs were a factor in their purchasing decisions.

Organizations with healthy cultures of maintenance have leaders who are true believers. In some cases, as with Fiix, the organization is founded by somebody who has a deep understanding of maintenance. But most leaders need to recognize the need for better maintenance through their own experience. One of our favorite examples comes from Widener University, where the president, Julie Wollman, moved into a temporary location on campus during renovations to her office. She learned firsthand about the poor condition of some spaces, and reflected on her experience: "Maintaining a campus well is a challenge, but everyone, in every building, should feel respected and welcomed on our campus by clean and well-maintained spaces. This requires our attention. Everyone's work area should be maintained as well as the president's suite."[10]

Over time, we have noticed that many examples we encounter hinge on an organization's ability to change its routine behaviors. Changes along these lines are most sustainable when they're backed by leadership. Once leaders are convinced to pay serious attention to deferred maintenance, they have essentially taken the most difficult step.

MAINTENANCE REQUIRES CONSTANT CARE

The final principle of the maintenance mindset is that maintenance requires constant care. Maintainers are most effective when they can focus on their work, improve and refine their methods, and apply their innate curiosity and ingenuity. This has been true throughout the history of maintenance, whether you're talking about the development of better tools, forums where experts can exchange ideas, and more recently, digital spreadsheets and applications to keep track of it all. Although we are incurably skeptical of claims that innovation will solve all problems, we think that maintenance work is one area where *actual* innovation has paid off, and where further innovation is necessary to accomplish the goal of keeping things in good working order.

In chapter 3, we described how the development of computerized maintenance management systems (CMMS) in the 1980s—first "home grown" within companies, then eventually created and sold by third-party vendors—marked an important shift in how maintenance work is performed and managed. The adoption and improvement of these systems demonstrates the benefits of constant care.

At its core, a CMMS is a database that keeps track of equipment and tasks. It has interfaces that help users make decisions (such as scheduling and purchasing), and schedules that keep records about the tasks associated with maintenance (such as budgeting, purchasing, uptime/downtime, work orders, "wrench time," and so on). It's important to note that a CMMS is like any other database: a powerful and flexible tool whose impact in the world depends entirely on how it is implemented, managed, and used.

This combination—new digital tools to tackle problems that are costly and ubiquitous—makes for enticing investment potential. One recent report from the market research company QYR measured the global CMMS market size at $787 million in 2018, and predicted that it would double by 2025.[11] This growth boils down to the fact that the workforce is changing—getting more digital savvy and digital dependent, more mobile, and more diverse—and CMMS companies are working hard to meet these needs.

A good example comes from UpKeep, a California-based company that was founded in 2014 by Ryan Chan. Ryan, a chemical engineering graduate from Berkeley, was working on a team charged with reducing downtime for equipment in a water treatment plant. Ryan searched around and realized there was no product on the market that did what he had in mind, so he decided to build it himself. His key insight was that maintenance staff don't sit behind desks all day. As they move throughout the field or shop floor, they usually carry smartphones or iPads. But existing maintenance software wasn't designed to work well on mobile devices. As a result, maintenance staff would have to take notes on paper and then walk back to their desks to enter them into the CMMS. Ryan's idea for a "mobile-first" maintenance app resonated with people working in small teams who needed to share information about breakdowns quickly. UpKeep built some traditional CMMS functions around this mobile-first approach: work orders, workflow, and data analysis that could help users move from reactive to preventive to predictive maintenance.

Ryan's quick mind and infectious optimism helped to earn the support of Y Combinator, the famous California incubator. In 2018, UpKeep went on to raise more than $10 million in series A funding from Emergence Capital, which saw tremendous potential in UpKeep's ability to reach the "deskless workforce." Ryan tells us that the typical path for finding new customers starts when everyday facilities managers or maintenance staff discover the app, find it useful, and then become UpKeep's champions in promoting it to management.

Another example of a forward-looking—even innovative—company in the CMMS space is Augury, a start-up founded by Saar Yoskovitz and Gal Shaul. Saar and Gal were classmates at Technion, Israel's premier research university, where they both aspired to start their own company. They shared an apartment after graduation and soon discovered an opportunity to use both of their talents in a field they refer to as "machine health." Gal was working as a software developer at a medical device company and visited a customer to understand why a machine wasn't working. As he entered the room where the machine was, he could immediately hear that the fan was all clogged up. So Gal, a software developer, found himself solving the problem by washing the filter.

As Saar remembers it, Gal came home that night and said, "'I can hear something is wrong: Why can't my code make my computer understand that?'" Saar was proficient in audio analysis, so the pair was soon exploring how to diagnose machines based on sound. Saar remarked, "It's something that is very natural for human beings in general. You hear the car. You hear the refrigerator. But for some reason we didn't know of any solution that does that automatically."

Saar and Gal spent the next few months investigating different machines in different market segments—trucking, temperature-controlled shipping containers, and other areas where machine health was mission critical. They soon realized that they could focus on commercial real estate, such as office buildings and shopping plazas, where the similarities across buildings would provide a basis to develop a more complex algorithm for predicting machine health. As Yoskovitz put it, "Every factory is a bit different than its neighbor. They have customized machines and they use them differently. Whereas at a building, a pump is a pump and you have tens of millions of them out there. It's a much less complex machine than a big production process machine in a factory. So we decided to start there."

Their first product involved placing sensors on pumps, collecting and analyzing the data, and returning the results rapidly so that users could know if they needed to replace a bearing or balance a motor. This became a signature of Augury's approach, which bundles together three different technologies: sensor connectivity, machine learning on the company's servers, and real-time diagnostics. In other words, Augury is neither just a software company nor just a system for tracking work orders and inventories; rather it combines the traditional function of a CMMS with hardware sensors and predictive analytics.

The approach put them on the cutting edge of a market where industrial giants such as GE have invested increasing amounts of resources: the "Internet of things," or IoT, a world where any device—baby monitor, refrigerator, car, and so on—can be accessed via the Internet. One market research firm—specifically citing the ability of predictive maintenance to reduce downtime and improve safety—

estimates that the industrial Internet of Things will be a $950 billion market by the year 2025.[12] With the vast amounts of data generated by IoT, companies will have opportunities to provide new types of feedback and intelligence to customers—provided they can find some meaning in the sea of data.[13]

Even if we cast aside our skepticism for the innovation-speak that is invading the realm of predictive maintenance—don't even get us started on maintenance and the "fourth industrial revolution"[14]—it is clear that innovative approaches to maintenance will continue to grow. Users are seeking cloud-based, mobile-enabled, user-friendly, flexible, and data-driven solutions for all kinds of maintenance needs. They want a more reliable, sustainable, and well-maintained world. But technology alone won't make it happen; it can only succeed through diligence, attention, and the kind of creativity that moved Saar and Gal to write software that could "hear" a clogged fan.

TAKING A LONG VIEW

Although we've grown tired of being immersed in a culture rife with the Innovation Delusion, we feel optimistic that people in all walks of life—from blue-collar workers to well-compensated executives—appreciate the need to take a longer view. For example, the Long Now Foundation was founded in 1996 to "help make long-term thinking more common." Based in San Francisco, its leaders include some leading lights of digital technologies, including its cofounder Stewart Brand, computing pioneer Danny Hillis, and *Wired* cofounder Kevin Kelly. There are signs of a movement toward long-term thinking visible at the heart of mainstream American capitalism as well. In 2019, the Business Roundtable—an influential organization of CEOs of American corporations—announced its support for corporations to move away from quarterly earnings reports. In a letter signed by 181 CEOs, they declared, "Public companies should be managed for long-term prosperity, not to meet the latest forecast."[15]

We get energized when we think about the maintenance mindset because it encompasses so many things that are so important to us: connecting with people, supporting their goals, caring about their experience, and working together. The maintenance mindset is, in many ways, the antidote to the Innovation Delusion. The Innovation Delusion is the false belief that the pursuit of innovation and novelty will lead us into the promised land of growth and profit when, in reality, it will lead us to ignore the ever-accumulating pile of deferred maintenance and infrastructural debt—and, in the process, lead individuals toward burnout and our society to accelerating levels of exploitation and inequality.

By contrast, the maintenance mindset starts with the recognition that maintenance sustains success—as long as you can identify what you believe to be good, and what you want to sustain. To accomplish this requires awareness that maintenance is an investment, not a cost, and should be recognized and supported accordingly. Finally, embracing maintenance does not mean turning your back on innovation; instead, it encourages you to find places where innovation and improvement can be useful tools to support your core values and connect to a deeper purpose. In the concluding chapters of the book, we'll see more examples of individuals and institutions who are using a maintenance mindset to repair the damage done by a society obsessed with quick fixes.

CHAPTER NINE

Fix It First

REPAIRING OUR BROKEN INFRASTRUCTURE

When Rudy Chow became the head of Baltimore's Department of Public Works in 2013 and assumed responsibility for the city's water system, he faced what he described as the "modern day utility nightmare challenge." The problem wasn't the water system itself. He's said that the city has "an awesome, awesome system." For instance, the network of pipes was built with lots of redundancies, so if one pipe had to be shut down, they could still move water around that section and get it to customers. The issue the city faced was decades upon decades of deferred maintenance. "The sixties, seventies, eighties, and even nineties were the big buildup period in the United States," Chow told us. "A lot of funding went to system expansion. Very little thought was given to what we should do with pipelines that were already underground." The result was deferred maintenance on a massive scale. Chow is quick to point out that Baltimore is not alone. "As I talk to my colleagues across the country, we all face the same thing."

Unlike many people who come to lead public works departments, Chow isn't a politician. He passionately identifies himself as an engineer, which entails a focus on rational, anticipatory planning. "In the utility world, there are two schools of thought," Chow told us. "One is basically 'fix it when it breaks.' The other is proactive—'let's try to get in front of the curve, let's try to anticipate failure rates and prioritize so we can prevent failure from happening.' " If it's not clear from that description, Chow is a firm believer in the latter approach.

For Chow, getting on top of deferred maintenance and crumbling pipes required a series of difficult decisions. The most difficult of all were a string of rate hikes to bring revenue in line with costs, including the looming price of deferred maintenance. As Chow told us, "We recognize that our citizens hate water rate increases, but as professionals, we cannot just ignore [reality]. Somebody had to be an adult here. And we choose to be that even though we know we're going to get hit hard. But we believe it's the right thing to do."

Chow retired in February 2020. One legacy of his tenure as director was the effort to change the culture at the Department of Public Works in deep ways. "I found myself in the position where we were in a one hundred percent reactionary [or repair-centered] posture, so now I'm trying to transform that culture from a reactionary culture to a proactive culture," he said. His guiding philosophy is asset management, a systematic approach to monitoring, caring for, and planning resources. Among other structures, Baltimore's system has about four thousand miles of water lines, twenty-five hundred miles of sewer lines, and twelve hundred miles of storm water lines. Replacement costs about $2 million per mile, and the total cost to replace the system would be about $15 billion. Chow explained, "From a strategy perspective, we know we don't have fifteen billion dollars and it makes no sense to replace them all at once. We want to do a proactive strategy where we replace X percent a year, and then when we finish, we start all over again."

Overall Chow and the other administrators at the Baltimore DPW see reasons for hope. For the past eight years, the city has experienced double-digit rate hikes nearly every year. But when Chow and his staff project costs "seven, eight, ten years out" they see water rate increases "coming down to just slightly above inflation." To Chow's engineer mind, this future is the dreamed-of "steady state," where deferred maintenance has been addressed and the overall system is working well. "We are winning," Chow told us. "We are putting our city on a very good, solid foundation and good footing. If we just maintain our course, we can stay ahead of these failures and these emergencies."

Cities are chaotic places. In the twelve months before Chow's retirement, the city's mayor, Catherine Pugh, resigned under the shadow of a corruption scandal in early 2019, and Baltimore's computer systems were taken over by a ransomware attack, leading the Department of Public Works to delay sending out water bills by more than three months. Maintainers must grapple with reality, which is complex, difficult, and uncertain even at the best of times, and for many American cities, these are far from the best of times.

Moreover, in late 2019, Chow and the DPW faced fierce criticism from residents and the city council for raising rates for years and not being responsive to complaints from citizens. White residents have disproportionately moved out of the city over the last four decades, which means the costs fall unduly on the shoulders of African American residents, who are often poor. "Some families have to spend up to 8% of their income on water bills, which is a really significant amount," Coty Montag of Baltimore's NAACP Legal Defense Fund told one reporter.[1] The question of how to pay—and *who* should pay—for deferred maintenance is an unanswered question of justice in our country.

Our point in recounting Rudy Chow's story is not to present his work as a panacea. You don't run an infrastructure agency and remain free from criticism. Nor are we claiming that Baltimore's Department of Public Works has it all figured out. Rather, the steps Chow took as director illustrate what we described in the last chapter as the *maintenance mindset*. Just as organizations can benefit when their members adopt a maintenance mindset, American infrastructure policy would profit from approaches that put maintenance front and center when it comes to both dealing with existing systems and planning new ones.

The two first steps of adopting the maintenance mindset involve coming to grips—often painfully—with where we are at on deferred maintenance and then starting to think about maintenance costs ahead of time. As we'll see, both of these steps—but especially the first one—face real obstacles: Oftentimes, we simply lack knowledge and measurements of the conditions of our infrastructural systems, which systems need attention first, how much their repair or replacement will cost, and so on. So, in order to even begin the process we have to get up to speed. The typical picture we discover is frankly depressing, but as Rudy Chow and others you'll meet in this chapter show us, success is attainable and the future is not without hope.

MEASURING THE PROBLEM

As we saw in chapter 4, for decades the term "infrastructure policy" has entailed building new things rather than keeping up existing things. But experts on the policies and politics of infrastructure have increasingly seen the error of this approach. As Rick Geddes, the director of the Cornell Program in Infrastructure Policy, told one reporter, "Our challenge in the 21st century is not so much building out new extensive networks of roads, it's taking better care of what we already have."[2]

People who have made this shift sometimes rally around the motto "Fix It First." There are varying opinions about how much new infrastructure we need, but the core idea is that Americans should generally prioritize maintaining and repairing existing systems before they erect new ones.

However, digging down into precisely where we should begin rehabilitating infrastructure leads to tricky questions. Part of the issue is a lack of standardized information about the state of individual pieces of infrastructure. In its 2016 report "Bridging the Gap Together: A New Model to Modernize U.S. Infrastructure," the Bipartisan Policy Center—a think tank focused on taking good ideas from both major parties—recommends "establishing a standardized inventory of the physical and economic condition of all public assets," including "transportation infrastructure (streets, bridges, stations, ports), water systems, civil buildings (schools, courthouses, convention centers), vacant land, and underutilized real estate."[3] The center argues that measures should include the cost of maintaining each asset "over its remaining useful life, the cost of replacement, and the potential impact of a failure."[4] These kinds of measures would enable political leaders to create policies and prioritize spending based on the actual needs of the structures.

For most localities, doing such measurements would be sobering. A big reason the true cost of infrastructure and the maintenance thereof is not visible is a trick of accounting. Municipalities are not required to count infrastructure as liabilities, even though they are on the hook for taking care of them in perpetuity. Many infrastructure advocates are pushing to change this, including Chuck Marohn of Strong Towns and members of the Bipartisan Policy Center. As we saw earlier, Marohn and some colleagues estimated that the city of Lafayette, Louisiana, had infrastructure needs of about $32 billion but a tax base of only $16 billion. His research and travels around the country have convinced Marohn that Lafayette's finances are the norm across America.

Making this accounting shift would be painful, pushing the books of most American cities massively into the red, but it would provide a more realistic picture of where we are and would allow us to grapple with reality, even if only on a triage basis. Jill Eicher of the Bipartisan Policy Center likened the accounting shift "to the dawning that took place about unfunded public pension obligations."[5] Governments didn't include pensions as liabilities until an accounting rule required them to do so in 2012. Counting pensions as liabilities had staggering consequences, and came at a time when many state and city governments, including Detroit and Flint, Michigan, faced bankruptcy—in part because of pension plans. But counting those costs was simply the honest thing to do. The same is true for infrastructure.

There are limits to these kinds of quantitative approaches. For one thing, infrastructure is highly varied and it may be difficult to come up with standard measurements that cover all of it. For example, we might have to choose between maintaining roads, dams, and schools. How do we meaningfully compare their contributions to our lives and the state of their degradation? There is also often a significant gap between how much something costs in financial terms and how people *value* it in moral terms. Community members might value a road because it has scenic overlooks and access to good fishing holes, even though the road doesn't contribute to the local economy and will be expensive to keep up. Moreover, as the historian Jerry Muller shows in *The Tyranny of Metrics,* and others have argued elsewhere, quantitative measures can create perverse behavior, leading individuals to become more concerned with getting good "grades" than with taking care of the things the numbers are supposed to measure. Finally, we should simply recognize how infrastructure spending has traditionally been driven by the politics of pork, with legislators trading favors to bring money and projects back home. Still, we agree with other infrastructure advocates that we can do a lot better at counting infrastructure. Such numbers could allow us to put politicians on the hook for taking care of existing infrastructure, not just building new stuff.

LOOKING DOWN THE ROAD

If you ask Chuck Marohn what he recommends localities do about infrastructure, he gets quiet. A big part of his philosophy is that there are no premade cure-all solutions that can be applied in all situations. His general recommendation is that planners and citizens start by paying attention to small details in their communities. Often this means literally walking around by climbing out of our cars (which is the medium through which we often experience small to midsize towns in this country) and getting a *feel* for the place by examining which neighborhoods are thriving and why, how different parts of the community are or are not connected, and how infrastructure like roads is contributing to this picture.

Marohn gave us a hypothetical example of a neighborhood in his town where less well-to-do families have been cut off from grocery stores and other necessary shopping and civic spaces. Should the city build an attractive footpath to compensate for the lack of sidewalks to and from the neighborhood? Should they create a new bus line? Offer incentives for someone to open a small grocery store within the neighborhood? No solution is perfect and every decision, including the decision not to act, has costs, but it is only by attending to such local, internal details—as opposed to, say, erecting yet another strip mall on the edge of town—that you can hope to benefit the community.

But Marohn's biggest recommendation, and the idea running throughout most of the Strong Towns message, is one of caution. Communities should think twice—or, more like a dozen times—before building new infrastructure, and especially before taking federal money to do it. Federal funding to build structures often comes free or via cheap loans, but communities are adding to their debt load by taking that money and building new stuff.

Today, it's very difficult, sometimes impossible, to know if a new piece of infrastructure is worth the downstream cost. For instance, one of us lives in a town that recently replaced an intersection with traffic lights on a busy stretch of state highway with a partial cloverleaf interchange. Honestly, the new interchange is pretty nice. It's visually appealing; it cuts down on traffic by ensuring no one needs to stop on the highway; and it likely makes that stretch of road safer. But are those benefits worth the downstream costs? After all, in twenty years or so, that interchange, which cost millions of dollars to construct, is going to be looking mighty shabby and require expensive rehabilitation. Eventually, it will need to be replaced altogether. Is it worth it? Right now, it's impossible to say.

Marohn believes this lack of knowledge is the norm across the nation and is reflected in how we've failed to think through the costs of convenience. In an early series of blog posts, Marohn recounted the decades-long history of how small, often rural roads lined with houses were connected to larger arteries and eventually highways with interchanges. Doing some basic estimates with a twelve-mile stretch of road, Marohn found that all of this additional infrastructure shaves a minute and thirty-five seconds off the trip but adds millions of dollars to our infrastructural debt burden.[6] The true benefits and costs of this enormous shift in how we do things—the great post–World War II experiment in organizing our society—have never been weighed.

A ROLE FOR THE FEDERAL GOVERNMENT

Chuck Marohn has a libertarian bent, and he sees little positive role for the federal government in improving America's infrastructure problems. When we asked him about iconic examples of American infrastructure born of federal programs—such as Hoover Dam or the Interstate Highway System—he said that he wouldn't have signed on to any of those programs if he was alive when they were undertaken. When discussing Hoover Dam, Marohn pointed to the environmentalist Marc Reisner's book *Cadillac Desert*. In that book, Reisner concludes, "If history is any guide, the odds that we can sustain" the infrastructurally intense water culture of the American West "would have to be regarded as low."[7] In Marohn's view, it's a classic example of an unhealthy mix of federal power and private interests coming together to build something unsustainable.

But you don't have to accept Marohn's conclusions about the federal government to learn from his thinking. We believe that all levels of government can play a more positive role in finding solutions to our current problems. But we believe that *much more federal infrastructure spending should go to maintenance, not building new systems*. Indeed, we think that, for the time being, "Fix It First" should be the reigning motto of our national infrastructure policy.

And here we mean *true* maintenance—conserving what we already have. Sadly, in the infrastructure policy world, "maintenance" is often a euphemism for upgrades that widen and expand a roadway and add new traffic and pedestrian control systems. Some of these changes can be good, but all of them add technology debt. We need to clarify when we are simply maintaining systems and when we are adding to them in ways that increase our overall burden. (To be clear, there are important exceptions to our emphasis on true maintenance. We believe in upgrading infrastructure so that it is truly accessible to people with disabilities. Making these changes will be expensive, but it is just the right thing to do.)

There's a role for new construction, too. When we talked to Casey Dinges, one of the American Society of Civil Engineers staff members who helped create the Infrastructure Report Card, he pointed out that the United States is still a growing nation, with a population increasing at about 1 percent a year. At that rate the population doubles every seventy years. How do we prepare for that eventual reality? Some new projects would no doubt contribute to economic growth and human well-being—though such projects should be planned and undertaken with more wisdom than the pork barrel politics that drive the status quo.

There are legitimate arguments today about how much federal debt really matters and whether the current U.S. debt load is even a problem. In some visions, we should actually be doing more federal spending on infrastructure. But even if you think we should spend more bringing American infrastructural systems up to code, federal spending alone simply isn't a sustainable answer to future developments, if those developments mean that state and local bodies will have to maintain those structures for the rest of time. Americans need to make wiser choices.

LEARNING FROM OTHERS

When the historian of technology David Edgerton—who is one of the leading thinkers in the study of maintenance history—visited us a few years ago, his talk turned to the sad state of American infrastructure and particularly the famously ugly, run-down appearance of New York City's Penn Station. What explained the gap between American rail stations and their better maintained and attractive European counterparts? "Hmm," Edgerton said, pausing for a moment, "Well, have you ever heard of civic pride?"

Edgerton may have been ribbing us, but his joke contains a germ of truth. As we have seen repeatedly, standards of maintenance and order have changed a great deal over time. They are culturally dependent. When you ask people who think a lot about infrastructure if there are examples of cultures or nations that are good at maintenance, they bring up a few repeatedly.

One is the Shinkansen, a high-speed rail system in Japan that began operations in 1964. The Shinkansen is a marvel of efficiency and safety, in large part because of its highly developed maintenance practices. No one has ever been killed by an accident on the Shinkansen, and there have only been two derailments in the system's history—one from an earthquake, another from a blizzard. (For contrast, check out the sprawling Wikipedia page "List of accidents on Amtrak.") With a top operating speed of two hundred miles per hour, the line has had an average delay of less than one minute per train, with the exception of 1990, when it just surpassed that mark. In 2002, the average delay was twenty-two seconds.[8]

Tell that to someone who uses Amtrak trains, and you may see them break down in tears of frustration. Only 75 percent of the trips on Amtrak's highly trafficked Northeast Regional run on time. Cross-country lines, like the Empire Builder and the California Zephyr, have an on-time rate as low as 20.9 percent.[9] Amtrak's delays are attributable in part to aging infrastructure and systems. As of January 2019, Amtrak's railcar fleet had an average age of thirty-three years,[10] and its trains have to run slower along some stretches of their journeys because those portions of track are so degraded. Some spots on Amtrak's lines are famously rough, leading to a bumpy, spine-jolting ride for passengers.

All of this can be contrasted with the Shinkansen. The Japanese rail line takes pride in its maintenance practices—most notably the inspection trains known as Doctor Yellow. Named for their bright yellow paint jobs, the trains house semiautomated systems and monitoring equipment that records track conditions in fine detail. The trains are famous among passengers and viewed as the key to a healthy rail line. Train watchers believe seeing a Doctor Yellow brings good luck, and they avidly share information about the railcars' whereabouts online.[11]

One thing that makes the Shinkansen so successful is high standards. Imagine if Amtrak worried about whether or not its trains were under a minute late on average. But high standards alone are not enough. Good public infrastructure also requires collective values, which is something evident when we look at another famous example of well-maintained infrastructure: Dutch flood prevention systems.

In 1953, the Netherlands experienced a devastating flood, which led to the construction of the famed Delta Works, a large interconnected system of dams, dykes, sluices, levees, and other barriers, which the American Society of Civil Engineers has described as one of the Seven Wonders of the Modern World. This system has stood the test of time, and though the entire country is located below sea level, the Netherlands hasn't suffered a major flood disaster since the Delta Works were built. The Dutch have become internationally recognized experts in water management, making their know-how in this area a centerpiece of their foreign aid efforts.

When Hurricane Katrina unveiled the shoddiness of New Orleans's levy system, for instance, architects, city planners, and politicians from Louisiana went to the Netherlands to see what they could learn.[12] And a New Orleans architecture firm, Waggonner & Ball, held a series of workshops pairing Dutch water experts with public officials in Louisiana in what became known as the Dutch Dialogues.[13] The health of Dutch water networks stands in stark contrast with degraded water control systems in the United States. For example, the 2017 ASCE Infrastructure Report Card gave American levees a D and reported that they would require $80 billion over the next decade to be brought up to satisfactory condition.

Dutch water management expertise exists because the nation decided to get on top of flood control and ensure it never again saw a destructive deluge like the one in 1953, but this decision and the habit of continually funding and maintaining such controls arises from a deeply held collective value: that water management is a communal responsibility that requires taxation. Local water boards, some of which go back to the thirteenth century, are the oldest form of representative government in the Netherlands, and the buildings that house these boards are a point of local pride, often emblazoned with colorful coats of arms.

We can get a sense of these collective values by considering a popular myth in which Dutch dykes play a part. In 1865, the American author Mary Mapes Dodge published *Hans Brinker, or The Silver Skates,* which contained a story known as "The Hero of Haarlem." In that story, a little Dutch boy discovers a leak in a dyke and saves the day by sticking his finger in the hole. The story has become a part of American folklore and is apparently not much loved by the Dutch. Maybe it's because people in the Netherlands realize that a dyke is an enormous, dunelike mound rather than a wall that could be saved by a finger. A statue of the little Dutch boy outside of the Dutch city of Spaarndam gets things more right. A plaque there says the statue "symbolizes the perpetual struggle of Holland against the water." It's something Dodge emphasizes in her story, too, though the lesson is often missed. "That little boy represents the spirit of the whole country," she wrote. "Not a leak can show itself anywhere . . . that a million fingers are not ready to stop it, at any cost." Dutch infrastructure, including its constant maintenance, is an achievement of the people in that they collectively back national and local policies, including taxation, that make the control systems possible.

Talk of communal values might feel unnecessarily touchy-feely, but the point is imminently practical. For example, experts in infrastructure policy have constantly called for a need to raise the federal gas tax, which was last increased in 1993 and has remained flat even though inflation rose 68 percent between 1993 and 2017.[14] Having a sensible gas tax is as basic as it gets, but so far it has remained out of reach in American politics. Again, we need—collectively—to face reality.

Infrastructure problems and lack of maintenance are not uniquely American. The Morandi Bridge in Genoa, Italy, collapsed in 2018, killing forty-three people and leading media outlets like *The Economist* to point out that infrastructural degradation wasn't only an American problem.[15] Meanwhile, observers asserted that Brexit was undermining British systems, and UK policy makers worried about a "catastrophic collapse in the nation's infrastructure."[16]

Such problems are even starker within poor nations around the globe. For instance, the famed development economist Albert Hirschman wrote in his 1958 book *The Strategy of Economic Development* that "the lack of proper maintenance" was "one of the most characteristic failings of underdeveloped countries and one that is spread over the whole economic landscape." He warned that infrastructural investments would come to naught unless nations developed a "maintenance habit" and a "compulsion to maintain." But the example of Dutch history shows that we can learn from our mistakes and develop the habits necessary to improve the human-built world that surrounds us.

SOMETIMES CHEAP SOLUTIONS ARE THE RIGHT SOLUTIONS

Many politicians cannot resist the bait of shiny, expensive infrastructure projects that are of questionable value, and the elected officials in New York are no exception. As a solution to LaGuardia Airport's transportation woes, Governor Andrew Cuomo has been pushing a multibillion-dollar "AirTrain," which transit advocates have pointed out would provide no improvement over existing modes of transportation and others have described as "lunacy" and as a "Port Authority turkey."[17] Meanwhile, New York City mayor and briefly presidential candidate Bill de Blasio has been advocating for a $2.7 billion streetcar known as the Brooklyn-Queens Connector, or BQX.

Public transit experts and advocates, however, assert that cheaper solutions are often more effective for providing what people need. Tabitha Decker, the deputy executive director of TransitCenter, a public transit advocacy organization, points to buses as one place where a lot of improvements can be made—and, in New York and several other cities, look like they *will* be made. "Essentially, we saw that there was a real problem. You've got more than a million rides taken daily on the buses in New York, but those buses were very slow and very unreliable," Decker told us.

TransitCenter made a revolutionary discovery in one study. It found that the two most important things to transit riders were "service frequency and travel time."[18] In other words, users want a system that works well and is reliable. It also found that "riders say the least important improvements are power outlets and Wi-Fi" out of a list of twelve potential improvements. As TransitCenter's report put it, "Our findings call into question the fad among transit agencies touting free Wi-Fi for customers who don't care strongly for it."

In 2015, the staff at TransitCenter decided to begin working on the bus issue. With a coalition of partners, it launched the campaign known as Turnaround by July 2016. The initiative focused on a few simple recommendations: modifying routes to reflect actual use and need; allowing riders to board a bus through all of its doors, not just the front one, and thus speeding up boarding; adopting better systems to keep buses on track and adequately spaced from one another; and redesigning streets where possible to make travel more efficient (for instance, by adding bus lanes).

With the recommendations in place, TransitCenter focused on a few basic strategies for getting these changes made. First, TransitCenter communicated and coordinated with staff within transit agencies, "especially those who would be the ones to actually implement the changes we're seeking," as Decker put it. The advocates and agency staff members didn't always agree. But as Decker said, "We respect the expertise and power that agency planners have and seek to build the ties with them and remove obstacles in their way." This is the opposite of seeing career staff members as "laggards" who need to be "disrupted."

Second, TransitCenter worked hard to keep the issue in the media by putting out information that was newsworthy. For example, the agency started releasing bus arrival data so that app developers could create apps that enabled riders to know when their bus was arriving. TransitCenter harnessed and repurposed that data to put out service report cards for every bus in the system. "Never before could a bus rider, an elected official, or a journalist take a look and see exactly how slow and unreliable our buses are," Decker explained.

The Turnaround coalition also involved transit unions, including bus drivers, to push for riders to be able to enter all doors. Drivers were happy to see this happen because asking for fares was the situation that created the most tension and led to the most violence on buses. In the all-door entry approach, bus drivers are no longer responsible for collecting fares.

The group also held a number of attention-getting events. Coalition partners rolled out a giant red carpet in front of city hall, representing a bus lane, and the bus riders' arm of that grassroots group walked a fake bus down the red carpet. "TV news loves that kind of cheesy shit," Decker explained. On a more serious note, the group noticed how, from his first mayoral campaign onward, Bill de Blasio said that his goal was to make New York a "fairer city." TransitCenter and its partners emphasized that bus riders are, as Decker outlined, "more likely to be lower income and more likely to be people of color, and the fact that we, as a city, weren't improving the service was having the strongest impact on those people." And in 2018, the coalition released its report titled "Fast Bus, Fair City," which echoed the mayor's stated goal.[19]

The efforts of TransitCenter and its partners have paid off. When Andy Byford, who resigned in January 2020, became the new head of the New York City Transit Authority and showed up on his first day, he announced a few priorities. Among them was retooling the city's buses and making the New York transit system more accessible to people with disabilities, another thing the TransitCenter has been advocating for. When the city released its bus plan, it was essentially the Turnaround coalition's agenda plus some other initiatives. It had taken about a year and a half of intensive advocacy.

Decker is an optimist of sorts. She points out that other transit advocacy groups have seen success in cities like Chicago and Miami. What stands out to us, though, is that these on-the-ground initiatives often involve making existing systems work better. No doubt systems require change, including upgrades to make them more accessible, but often the changes that affect most riders' lives involve rather cheap solutions, like improving bus service rather than pouring billions of dollars into the kinds of vanity projects favored by many politicians. Focusing on efficacy rather than glitz has the greatest impact. Again, often the most effective solutions are the least expensive ones.

We have noticed that some small towns in the United States—such as Tuscola, Illinois—once had asphalt streets but now have moved toward surfaces of macadam (also known as tar-and-chip), which is made by mixing crushed stones with tar. In terms of technological development, this is moving "backward": Macadam roads were popular in the late nineteenth and early twentieth centuries before asphalt became affordable and widely used. Macadam is not as attractive, and speaking plainly, it can be frigging annoying. Pebbles inevitably get unloosed from the surface and flung into your yard. But from a financial perspective it can make sense: Macadam is cheaper, at least up front. The question is, When does it make sense for a community to use a less attractive but cheaper solution? What is the value of financial solvency?

In some cases, the goal should be selective and graceful degrowth—paring back our infrastructural burden and getting smaller. People often bring up Detroit in conversations about the future of cities. There's no denying that the people of Detroit went through a tough time, and the place has often been the subject of "ruin porn" photography. But these days, the city is a happening place with new businesses, including "tech start-ups," trendy bars and restaurants, and, yes, even urban agriculture on the plots of land that once held houses—which is not to say all its problems have disappeared. The process of getting to this place, though, was extremely painful, most notably when the city went into financial insolvency. This raises the question of how we can help cities and rural towns shrink gracefully when their populations decline. One can imagine policies aimed explicitly toward this goal—including, for instance, more grants aimed at tearing down abandoned and collapsing buildings, which are a real problem in so-called Rust Belt regions and many rural towns.

REFORMING SICK GOVERNANCE

In some cases, significantly improving infrastructure will require political and governmental reform. Time and again, when you look at the transit systems of big cities, you find fractured, chaotic, governance structures: boards of directors that are simultaneously accountable to everyone and no one, elected officials overseeing systems they have little incentive to care for, and different modes of transportation being controlled by different transit agencies that poorly coordinate and sometimes even compete. The San Francisco Bay Area, for example, has twenty-six different transit agencies, leading to poor coordination and a dizzying series of choices for riders.[20] And the New York Metropolitan Transportation Authority (MTA), which includes the subway system, is famously overseen by the New York governor rather than the city's mayor. The mayor of New York City has traditionally been able to appoint only four of the MTA's twenty-one board members. This structure favors the suburbs, including by creating new capital projects rather than investing in the upkeep of existing urban systems.[21]

In 2013, the Regional Transportation Authority (RTA) in Chicago realized that it was struggling to effectively allocate funds among its various divisions. It hired a team of consultants, including the Eno Center for Transportation, to "help explore and identify best practices."[22] What the consultants' investigation uncovered, however, was that institutional barriers and poor governance structures were hampering the RTA. The Eno Center then teamed up with TransitCenter to do a wider study of governance in six major metropolitan regions: Chicago, Boston, Dallas/Fort Worth, Minneapolis/St. Paul, New York/New Jersey/Connecticut, and San Francisco Bay. The study found persistent issues in coordination and decision making, and made a series of recommendations: funding transit systems independently (rather than through annual appropriation packages that must be approved by legislators and governors), consolidating a region's transit agencies under a single administration, and making sure governing boards' leadership reflects the interests of core user groups and affected localities. Advocacy groups like TransitCenter often focus on what they can accomplish within current governance systems, because deep reforms can be hard and require huge expenditures of resources and political capital, but it can be difficult to imagine real change without these kinds of political transformations.

How do we put elected and appointed officials on the hook for maintaining existing infrastructure systems? We'll need to be creative. For instance, we've heard rumors of groups holding "ribbon cuttings" in front of decrepit New York subway stations to highlight just how run-down the city's structures are. We think this is yet another place where having better measures of the physical conditions and operating quality of infrastructure would be helpful. Up-to-date measures of how infrastructure is performing and what work it will need would enable advocates, opposing politicians, and ordinary citizens to hold officials' feet to the fire. Inversely, the measures would allow us to reward officials who are seeking the right solutions.

The more depressing reality is that our nation has no real answer for places like Lowndes County, Alabama, where a lack of infrastructure, including septic tanks, has led to hookworm infections and other public health problems. What do we collectively say to people who face such infrastructural nightmares? At the very least, liberals and conservatives should find ways to address issues where their respective philosophies and values overlap. To give one example, advocates focused on rural public health have argued that current government codes for septic tanks require expensive systems that are well out of reach for many poor families, basically guaranteeing that they will move to "straight pipe" solutions that pour raw sewage onto the ground or into local water sources like creeks. One solution would be to adopt cheaper septic systems that are used in poorer nations. These systems aren't up to code, but they are much better than nothing. As far as the codes are concerned in this case, the best has become the enemy of the good. Progressives should be in favor of reforming codes because it will benefit public health; conservatives should sign on because it is a clear example of government regulations holding back positive action.

Should we view sewerage, clean water, electricity, heat, healthcare, and access to a phone and perhaps even the Internet as human rights? In general, we believe the answer is a strong *yes*. (We recognize there are limits. If an individual or a family builds a new house with a well on *very rural* land with polluted water, do we owe it to them to provide clean water in some other way? Probably not. But this is a deep and unanswered question about where personal and societal responsibilities begin and end.) Moreover, as we saw with the Baltimore Department of Public Works example at the beginning of the chapter, dealing with deferred maintenance, for instance, by raising utility rates can profoundly undermine poor people to the point of losing their homes. How do we deal with this inequity? Treating basic infrastructure as a human right would require a profound shift in our culture. For this cultural sea change to succeed, we will all need to spend less time talking about impersonal "infrastructure" and more time talking about how these systems benefit and harm human life. The suffering in places like Baltimore; Flint, Michigan; Lowndes County, Alabama; and thousands of other urban, suburban, and rural places around our country is a testament to how far we have left to go.

CHAPTER TEN

Supporting the Work That Matters Most

MAKING THE WORK OF MAINTENANCE MORE SUSTAINABLE

Francia Reed deployed to Balad, Iraq, in early 2008. She had first enlisted in the U.S. Air Force more than thirty years earlier, in 1976, when she was trying to find a way to pay for college. Eventually, Francia earned bachelor's and master's degrees, and worked as a nurse in the labor and delivery units of private hospitals. Now she had decided to go back where it all began, rejoining the Air Force Reserve at the rank of captain and serving as a clinical nurse on the outskirts of a combat zone.

The hospital at Balad Air Base, about forty miles north of Baghdad, had teams of nurses whose jobs were to treat wounded soldiers, stabilize them, and prepare them for the next step in their recovery—usually putting them on a plane to another military hospital in the Middle East, Germany, or back in the United States. "It was a profound experience for me in patient care," Francia told us. "Having had the background as a maternity nurse, most of the time my patients were healthy. Most of the time it was a happy situation." But expectant mothers can be a handful, and Francia remembers a few patients who acted as though a nurse's job is "to wait on me hand and foot." She recalled, "They would put their call bell on, and I'd say,

'Can I help you, what do you need?' And they'd say, 'Can you hand me my water glass?'" Francia laughed, remembering her (internal) response: "Seriously? The woman next door is in pain, and I'm trying to help her, and you just needed me to move your water glass? You couldn't get that?"

The soldiers in Balad were different from the mothers in upstate New York. So were the social hierarchies. "In the army, they're very rank conscious. The enlisted soldiers didn't want me, a captain, doing things for them. I would tease them and say, 'But it makes me feel important. This is how I justify my existence here.'" Eventually, Francia would convince the soldiers to rest, and allow her to do her job. "I've never seen so many people so grateful for a glass of water and pain medicine. I don't know—it just made me put things in perspective."

We saw in chapter 6 how our society portrays maintenance work as low in both status and prestige. Hierarchies of occupational status are established and enforced in us from the time we are children. Through STEM camps and children's books, like Richard Scarry's *What Do People Do All Day?,* children learn that they should look up to innovators—Astronauts! Scientists!—and at the same time look down their noses at maintenance, care, and repair workers.

But there are different rules in a military hospital: The caste-like structures that keep maintainers at the bottom don't exist there. It's enough to make us wonder: What would the world be like if this status inversion were more widespread? What if everyone treated maintainers and care workers the same way that injured soldiers treated air force nurses—with deference and respect?

We'll meet several people like Francia in this chapter—people whose ideas and experiences can help us think about how to dismantle the caste-like structures that our society has built around maintainers. Some of their ideas will strike you as intuitive. It's a no-brainer to raise salaries for maintainers, for instance. But other concepts will be more complex or even controversial. It's important to capture this complexity, particularly since many of these occupations are usually taken for granted or dismissed as being "unskilled"—when, in fact, this work is foundational for society. It's important for all of us to hear, directly from maintainers, about how we could improve their working and living conditions.

ACCURATE PICTURES OF LIFE WITH TECHNOLOGY

One consistent theme in all of our conversations with maintainers, policy makers, managers, and executives is the need for better information for making decisions. For instance, there are few sources for comprehensive and reliable data about where different kinds of maintenance work is performed, and by whom. We lack Big Data, such as national or industry-level figures, for the economic value that maintenance generates; and we lack "small," qualitative data about how maintainers view and confront the challenges they face.

We laid out a puzzling contradiction in chapter 6. On the one hand, young people hear from universities, mainstream media, and many of their parents that they need to go to college, learn to code, and major in a STEM field if they want to get a good job and have a secure financial future. But this message starts to crumble upon closer scrutiny. For one, the pursuit of a four-year STEM degree makes a lot of students unhappy, and the odds are that they won't succeed. According to the American Society for Engineering Education, only 34 percent of engineering students graduate in four years. Deans of engineering schools often tell their freshman classes: "Look to your left, and look to your right. Only one of you three will be crossing the stage to get your diploma in four years."

There are indications that campaigns to drive more students into STEM degree programs might be doing more to advance the interests of universities and corporations than to help anxiety-ridden college students. In 2013, *IEEE Spectrum* published a series of articles under the heading "The STEM Crisis Is a Myth." In one eye-opening statistic, the number of STEM vacancies per year—277,000—was compared against the number of STEM degree recipients and H-1B visa holders—442,000. When annual supply exceeds annual demand to this degree, it should come as no surprise that there are an additional 11.4 million STEM degree holders who currently work outside of STEM fields.[1] Universities capture the tuition, and corporations benefit from oversupply in the labor market, since it increases competition and justifies keeping salaries and benefits low.

Now consider some fields in which the supply of qualified workers actually is insufficient. According to the labor market research firm Emsi, demand for workers in the skilled trades grew by 10 to 20 percent between 2013 and 2017. Examples include construction (12 percent growth, average wage of $19.18/hour), tile and marble setters (18 percent growth, $21.20/hour), and electrical installation and repair for transportation equipment (9 percent growth, $28.03/hour).[2]

While universities and policy makers continue to supply steady streams of revenue toward programs in innovation and entrepreneurship, there's ample evidence to suggest that society needs other skills—including emotional intelligence—much more urgently. One example comes from the medical fields, where the Bureau of Labor Statistics estimated that direct-care jobs such as personal care aide, home health aide, and nursing assistant were expected to grow by 26 percent between 2014 and 2024. Better social skills ensure appropriate levels of empathy in these jobs, but the economist David Deming points out other benefits as well: "Social skills reduce coordination costs, allowing workers to specialize and work together more efficiently." In a 2015 study, Deming found that jobs requiring high levels of social interaction grew as a share of the overall U.S. labor market by nearly 12 percent between 1980 and 2012. During the same period, jobs that were STEM-intensive but less social shrank by 3.3 percent.[3] Reporting on these trends, the writer Livia Gershon concluded that care work and emotional labor will only increase in importance as we continue into an era of postindustrial automation. Gershon noted that these skills are most typically found in jobs cast as "unskilled labor" and in working-class women, not in highly educated men.

Of course, data alone will not pave the path to more respect and better pay for people in the trades. Cultural myths accumulate over time, through multiple channels—and it takes just as long (if not longer) to dismantle them. This is one reason why we're grateful for people like Mike Rowe, host of the hit TV shows *Dirty Jobs* and *Somebody's Gotta Do It*. For years, Rowe has championed the dangerous, filthy lines of work that, as he describes it, "make civilized life possible for the rest of us"—from salt mining and shrimping, to sewage cleanup and sheep castration.

Rowe's easy demeanor and mischievous sense of humor might make it easy to miss the deeper and more profound message that motivates him. Rowe concluded his 2008 TED Talk with a plea: "The jobs we hope to make and the jobs we hope to create aren't going to stick unless they're jobs that people want. . . . The thing to do is to talk about a PR campaign for work: manual labor and skilled labor. Somebody needs to be out there, talking about the forgotten benefits."[4] This isn't a new idea. Echoes of Rowe's campaign can be seen in Studs Terkel's iconic book *Working,* and more recently in Senator Sherrod Brown's "Dignity of Work" tour.[5] To be sure, helping people find rewarding and meaningful work—even in dirty and unforgiving places—is a consistent theme in the American political and media landscapes.

The same theme is visible in the armed services, where maintenance and other forms of "dirty work" are essential for greasing the gears of the American military machine. One example comes from the U.S. Air Force, where high-status fighter pilots depend on low-status maintainers. It's not a stretch to say that air force strategists are obsessed with data and information. Aircraft and air bases embody massive investments, and there is tremendous pressure for the air force to ensure the longevity of those assets and investments. As a result, the air force has developed a sophisticated plan for measuring its maintenance needs, using advanced data and analytics to identify and fix a wide range of problems. The air force also happens to be one of the few organizations we have found that uses "maintainer" in actual job titles for the people who inspect and repair the electronic, mechanical, structural, and communications functions of aircraft.

In 2016, air force leaders realized that they faced a significant labor shortage. The air force had a nearly 16 percent vacancy rate for maintainers, four thousand short of being fully staffed for its sixty-seven thousand authorized positions. (By contrast, the air force needs thirteen thousand active-duty pilots to be fully staffed.) The financial and strategic consequences of the maintainer gap were significant, in light of the fact that nearly 30 percent of air force aircraft were not mission ready.

Commentators discussed a variety of reasons for the maintainer shortage. Some pointed to direct causes, such as budget cuts in 2014.[6] Others, such as the historian Layne Karafantis, highlight the unintended consequences of strategies that were much longer in the making. Early in the Cold War, air force leaders believed that automation would reduce the possibility of human error, and thus reduce the need for human workers. As Karafantis put it, the air force was "crudely trying to idiot-proof operations so that they could use whatever stock of personnel they could muster."[7] It's a good reminder that fantasies of automation are not unique to the early decades of the twenty-first century.

After discovering the maintainer shortfall, the air force set goals and created strategies to recruit four thousand new maintainers by 2019. It also produced some new recruiting materials to help it get there, such as "Maintainers: The Driving Force," a video released in 2016 that features clips of men and women covered in dirt and oil, turning wrenches, alternately grimacing and smiling, as a stirring musical arrangement builds in the background. The video begins with a voiceover: "We weren't the kind to go looking for the spotlight. We didn't need our hands held. We did our jobs just fine without getting pats on the head." This three-minute paean to maintainers features a few memorable lines: "The air force never did manage to build a plane that would fix itself." And, "Do the unsung heroes need to be sung about every now and then? Maybe. But don't sing too long because, with all due respect, we've got work to do." A line near the end nails the pride that maintainers share: "We are the men and women who keep this air force flying. We are the driving force."[8]

This recruitment campaign worked, more or less. In early 2019, Secretary of the Air Force Heather Wilson reported that the gap of four thousand maintainers had been closed.[9] But with one goal met, another problem was exposed: The air force was failing to retain skilled and experienced maintainers. A 2019 report from the Government Accountability Office cited an 8 percent drop in reenlistment rates between 2011 and 2017, including a first-time reenlistment rate of only 58.3 percent in 2017. In the GAO's summary, "Participants stated that the lack of experienced maintainers has increased workloads and stress levels." In other words, the maintainer shortage was on the precipice of a death spiral. The shortage itself was becoming a significant factor in the departure of maintainers, the very people the air force needed to train the new recruits.[10]

What strategies might the air force use to make things better for maintainers? Some answers are obvious. According to the GAO report, "Maintainers cited better pay as a reason to transition from the Air Force to the commercial aviation industry. They also noted consistent schedules, 8-hour work days, and overtime pay as additional benefits." Air force maintainers have also taken to Facebook, Instagram, and websites to commiserate. Their activity on social media features a rich mix of burnout, joy, pride, and humor. They often push the boundaries of political correctness when they share memes about depression, false promises of recruiters, and the follies of air force leadership.[11]

There's a lot to be said about the data and anecdotes that maintainers provide in their own unfiltered voices. We encourage you to look them up on social media. You won't find anyone clamoring for simulators, Big Data, biometrics, or macro-innovation, that's for sure. Rather, what you will discover are suggestions and needs that fall into the categories we've observed across all groups of maintenance and care workers: the need for better material rewards (such as pay and benefits); better intangible rewards (such as recognition and respect); and suggestions for fighting burnout by creating more space for maintainers to revel in the intrinsic joy of their work. We'll take these in turn.

PAY THE MAINTAINERS

If we want to make things better for maintainers, improvements in salary, benefits, and job stability are the best places to start. To underscore a point we made in chapter 8: a commitment to a maintenance mindset means recognizing that maintenance is a core value, and then devoting sufficient resources to that value. Let's spend a few minutes understanding the obstacles to giving maintainers better material rewards.

Many of the most popular digital platforms are sustained by unhealthy labor models. Consider, for example, the thousands of low-wage workers who monitor social media networks for offensive content and are responsible for removing it. Their work includes reviewing footage of rapes and murders (such as the live-streamed Christchurch massacre in 2019) to training algorithms to distinguish between dogs and cats. Analysts use a variety of terms to refer to these workers—"code janitors," "commercial content moderators," "ghost workers," "microworkers"—which speak volumes about the status bestowed on this form of labor.

These workers don't give TED Talks, they don't have philanthropic foundations, and they don't play volleyball or ping-pong at a lush corporate campus. But when they are successful, their work helps to preserve the appearance of software, social media platforms, and digital infrastructure as smooth, impersonal, and automated. Their work, ironically, supports the tenuous credibility of pundits who claim that automation will eliminate jobs. A growing number of scholars are documenting these workers, their vital importance, and their relative neglect. In *Behind the Screen,* the scholar Sarah Roberts suggests that this exploitation has hidden costs, including "damaged workers and an even more frightening social media landscape." The former Facebook contractor Chris Gray, for example, sued the company in 2019, citing his job removing pornography, hate speech, executions, and bestiality that users had uploaded to Facebook as the cause of psychological trauma and post-traumatic stress disorder that afflicted him.[12]

Fortunately, there are a variety of proposals for how to change things for the better. There is a clear need for interventions from the outside, whether in the shape of regulations for safe working conditions, wages, benefits, and support for workers who endure emotional anguish as part of their duties cleansing social media posts. And, in echoes of the solidarity and collective responsibility that information maintainers are calling for in their own ranks, full-time employees at software and digital platform companies are mobilizing to support their vulnerable colleagues. Employees walked off the job in 2018 at Google, for example, where the temporary and contract workers outnumber full-time employees (121,000 to 102,000, according to a 2019 report in *The New York Times*). Among their list of demands was better treatment for contingent workers—the temps and contractors who make up the majority of Google's workforce. One executive at a temp company told the *Times* that the phenomenon is "creating a caste system inside companies."[13]

Digital platforms like Google and Facebook have been celebrated for their leadership in innovation. Many are lauded as models for other companies to imitate, and examples of the results that innovation should create. But when we listen to the people who work in the industry, we hear a mandate for a different set of policy priorities: Companies should account for the importance of maintainers, protect them from undue harm, and compensate them in a way that reflects their contributions to these enormously profitable systems.

EATING OUR YOUNG

Beyond the monetary and material rewards of compensation, benefits, and job security, maintainers and care workers consistently talk about the importance of intangible opportunities and rewards: more support, more recognition, more opportunities to defy the caste-like condescension that they unjustly endure.

To bring some of these intangible opportunities into clearer view, let's consider a crisis that faces the nursing profession. Nurses are often the first people who spring to mind when we hear the term "care work." They hold the line between health and sickness, between life and death. Nurses are the maintainers of life itself.

Nursing is a relatively well-compensated profession, with an average annual salary of more than $50,000. Yet there is widespread panic about a shortage of nurses, given ever-increasing demand in light of the aging population and expansion of the American healthcare system. In 2018, the Bureau of Labor Statistics projected that employment of registered nurses would grow 12 percent over the next ten years, far outpacing the average growth rate of 5 percent it projected for all occupations.

Why is there a nursing shortage? One reason, according to the American Association of Colleges of Nursing, is the accelerating retirement rate of the existing nursing workforce. More than 50 percent of all nurses today are age fifty or older. Not only does their mass retirement leave hospitals and medical facilities shorthanded—as we saw above with air force maintainers—but when the most experienced members of the workforce leave, they take accumulated knowledge and sensibility with them as well. These demographic shifts have significant impacts: Nurses report higher levels of stress and lower levels of job satisfaction due to insufficient staffing, and a wide range of studies report that nursing shortages can make it harder for patients to access care.[14]

The Innovation Delusion afflicts many organizations that are determined to capitalize on the nursing shortage. The multinational corporation Johnson & Johnson is running a "Campaign for Nursing's Future" that conspicuously features innovation-speak. For example, its website includes a "Nurses Innovate QuickFire Challenge"—an unfortunate metaphor, given the position of nurses at the bloody front lines of the American epidemic of gun violence.[15] Elsewhere, companies have marketed "social care" robots with names like "Pillo" and "Pepper" that can dispense pills, keep track of nutritional and exercise goals, simulate what a human might say in conversation, and, of course, monitor and report on patients through cameras that are always watching.[16]

We were relieved to learn that some nurses share our skepticism about the invasion of innovation-speak into their domain. Francia Reed, whom we met at the opening of this chapter, recalled reading the literature in the field of nursing education and being struck by the repeated argument that "we need more innovation. But 'innovation' was very loosely defined if defined at all." She began to wonder about the empirical basis supporting claims that nursing needs innovation: "Is that true? Is that the case? How do we know that there's a cause-and-effect relationship?"

Francia's dissertation research took shape from those questions. She established formal definitions, conducted studies of innovation in nursing education programs, and found two things that surprised her. First, students who were exposed to teaching strategies that program designers considered "innovative"—clickers to respond to questions embedded in PowerPoint lectures or interactive case studies, "as opposed to the whole old-fashioned 'sage on the stage' kind of thing"—thought that these pedagogical techniques were standard. This must have been awkward to break to the faculty who thought they were on the cutting edge.

Her second surprise came through an exercise that had nothing to do with innovation. Francia asked students in her research study to keep journals about their ideas for improving patient care. The students had some good ideas, but "when I asked them, 'What have you done with these ideas?,' none of them had shared any of their ideas with anybody that they worked with."

Francia went on to describe her belief that improvements in nursing—what we, in this book, have called actual innovation—require organizational cultures that value and create venues for communication. She thought back to her active-duty service in the air force during the 1970s. "At that time," she recalled, "the air force actually had a formal suggestion program. If you had an idea about how to solve a problem, there was a form you filled out. There was a centrally located suggestion box. And the added incentive was that if your suggestion was adopted and it saved the air force money, you got a monetary reward."

When we asked Francia about how things could be improved for nurses, we were surprised that she didn't advocate for higher salaries. "I think if you would have asked me twenty years ago, I would have said, you know, nurses aren't valued and it's reflected in what they're paid. But that's come around, that's been corrected. The last survey that we did with our [SUNY Polytechnic Institute] graduates to find out what their salaries are—they're pretty good."

The deeper problems she sees are cultural problems, and they come from within. "I think that nursing still has a problem with what we call eating our young." This idiom, which has been around for decades and explored in books like Cheryl Dellasega and Rebecca Volpe's award-winning *Toxic Nursing,* refers to well-documented cultures of bullying and overwork in the profession directed at new nurses who haven't yet paid their dues. "If I could wave a magic wand, I would fix that," Francia continued. "I would try to do something in which we promote a culture where we say to new nurses, 'It doesn't matter that you just graduated yesterday. You are valuable. You are valued. We want you here and we'll hear what you have to say. You have a voice here. We welcome your contributions and we welcome your questions and we will support you.'"

These needs—to be nurtured, protected, and supported—too easily get trampled in societies obsessed with efficiency and profitability. Fortunately, there are educators like Francia who understand how important these needs are. But how are these needs met in aspects of our economic life where concepts like caring, nurturing, and supporting are relatively scarce?

Odds are good that if we asked you to keep track of typical interactions that lack a feel of caring or nurturing, you'd mention some encounter with customer service. The field of customer service has a bad reputation—some of it deserved. We've all had stomach-churning experiences when trying to contact companies with questions or complaints. Some companies made major investments to locate call centers in countries where local dialects don't mesh well with American ears. Others have allocated capital to automated services, in the hopes that customers will find what they're looking for through push-button menus or services such as Julie, Amtrak's "virtual travel assistant."

Herein lies one of the most poignant ironies of the digital age. The promoters of software and digital technologies have long promised their benefits for enhancing community and connectivity—but in many cases these technologies are being used to *reduce* human contact. To resolve this tension, our society needs talented and empathetic people who understand the importance of connections, know how to make other people feel like their perspectives are valid, and are able to direct their frustrations and concerns down a productive path. Our digital systems and digital societies need maintenance and care in this vital area.

People like Camille Acey give us hope. Camille is the vice president for customer success at Nylas, a software start-up whose products connect and integrate data across email, calendar, and contacts applications. Camille's role is to lead the team of employees who field calls from users and customers of the software that Nylas provides.

Camille, being well aware of the image problems of customer service, has developed a holistic approach to make sure that Nylas's customers see her company in a different light. First is the basic recognition that customer service involves more than the mere conveyance of facts. "It's not just a matter of updating and sharing a document," Camille told us. "One thing I really believe in is customers as partners. We rely on them as much as they rely on us. We are here because they are there, so we need to be thinking about things that we are not sharing but that we can share." Her role is one where she connects with customers, understands their perspectives, helps them learn from other customers, and—crucially—also brings these insights to the software developers at Nylas who are improving on their existing products and building new ones. Her workflow typically involves video meetings with customers in order to build a strong rapport. More than a typical sales executive guided by ever-higher commissions, Camille sees herself as an advocate for people who aren't in the room, and a translator who can provide value when a discussion spans the boundaries of different companies. Overall, Camille summarizes her vision of customer interactions in terms of "empowerment, accountability, and education."

Camille's broad industry experience has helped her understand that the cool, smooth, detached veneer of successful "tech" companies is, in fact, a mirage. She described working with one prominent Silicon Valley firm as "Keystone Kops," invoking the image of clumsy and incompetent policemen featured in silent slapstick comedies in the early twentieth century alongside comedians like Charlie Chaplin and Fatty Arbuckle. "It actually gave me some solace," Camille explained. "I expected this well-oiled thing, but I quickly realized these guys don't know what they're doing. That actually calmed me down." Camille's advice is to "focus on communication and taking our time. Taking our time—that's a really crucial one. Taking our time is really at the center of maintenance. In New York, they shut down the L train for the weekend because they can't fix it any faster than that. This really runs against the Silicon Valley idea of 'move fast and break things.' But that's what it takes."

BURNOUT AND JOY

Camille's work brings us to a final pair of themes that frequently arise when we speak to maintainers: burnout and joy. They work so hard that they have nothing left to give, yet their work brings them joy. They love the satisfaction of a job well done, and there's no better feeling than helping someone out of a pinch—whether it's fixing a leaky pipe, bringing another pillow to a patient coming out of surgery, or making sure that the classroom projector is working when the students arrive.

Forgive us if this is too simple, but if we want to improve things for maintainers—particularly their morale and mental health—an obvious step in the right direction would be to reduce the burnout and enhance the joy. It's not a zero-sum equation, as we'll explain below. But the inescapable fact is that maintenance and care work will be more sustainable if maintainers feel rested and appreciated.

Let's return to the digital world to explore burnout and joy within communities that produce open-source software. In recent decades, IT systems throughout all sectors of the economy have become more reliant on open-source software such as the Linux operating system, Python programming language, and Mozilla's Firefox web browser. One distinctive aspect of open-source software is its reliance on volunteers. All open-source projects have one or more individuals who are designated as the "maintainer." These people are responsible for answering questions, acting on bug reports, responding to requests for new features, and overseeing updates to the source code of the project.

The experience can be overwhelming. Nolan Lawson is a maintainer for PouchDB, an open-source database that web developers use to build "offline-first" applications that are useful when people are connected to the Internet and keep working even if connectivity drops. Nolan's essay "What It Feels Like to Be an Open-Source Maintainer" begins with an imagined scenario: "Outside your door stands a line of a few hundred people. They are patiently waiting for you to answer their questions, complaints . . . and feature requests. You want to help all of them, but for now you're putting it off. Maybe you had a hard day at work, or you're tired, or you're just trying to enjoy a weekend with your family and friends."

But the line outside Nolan's door never goes away, and the work of attending to it is frustrating. Some people are "well-meaning enough," but their code is "a big unreadable mess." Others "spew out complaints about how your project wasted 2 hours of their life." For open-source maintainers, the specter of burnout is never far away: "After a while, you've gone through ten or twenty people like this. There are still more than a hundred waiting in line. But by now you're feeling exhausted; each person has either had a complaint, a question, or a request for enhancement."[17]

When maintainers meet at conferences or chat online, they often talk about burnout and share strategies for coping. Jess Frazelle, another open-source maintainer, feels that "the hardest part is dealing with people problems. It might be that jerk that opens issues and is super mean, demanding, and/or condescending. . . . A lot of maintaining is keeping everyone happy." The key to accomplishing this, while staying sane, is to develop self-awareness and practice self-care: "Take time off when you need it. People may push you to extremes and make you feel you need to respond right away but listen to your needs as well."[18]

Another maintainer, Jan Lehnardt, advocates a different approach: Stop caring. In a blog post on burnout, Jan wrote, "The only thing that lets me sleep at night is not caring about any of these things. I'll get to them eventually, some may fall between the cracks. It's not nice from a project or people perspective, but short of leaving the project and leaving it all behind, I found this to be the only way to make my personal Open Source maintainership sustainable."[19]

Yet another approach to coping with burnout involves cultivating connections on a more intimate and spiritual level. Ariya Hidayat writes that open-source maintenance "is similar to any kind of hobby projects: soothing and therapeutic." Henry Zhu also tries to keep things in perspective: "It isn't just about the code (like all things in life), it's the people that keep the project moving forward and alive and the community of users. . . . Even though it's been a struggle at times, I just feel blessed to have been able to take part in all of this."[20]

Open-source contributors frequently talk about their work in terms of *passion*. In his autobiography *Just for Fun,* the Linux creator Linus Torvalds wrote, "It's been well established that folks do their best work when they are driven by a passion. When they are having fun. This is as true for playwrights and sculptors and entrepreneurs as it is for software engineers. The open source model gives people the opportunity to live their passion. To have fun."[21]

The truth of Torvalds's observation, that passion and fun enable people to do their best work, is born out when we consider maintainers back in the analog world—the mechanics who keep our motors running. To understand more about their universe, we spoke to one of our favorite mechanic/academic friends, Juris Milestone. We first met Juris, who teaches anthropology at Temple University, at a conference on maintenance that we hosted in 2016. His story pulls together a few of the themes common among maintainers across different walks of life.

Juris was born in 1969. "I grew up basically poor, with my mother," he recalled. "She had this friend group who she hung out with. They were all just blue-collar hippies," most of whom had served in Vietnam. One member of this group, a mechanic who had fixed jeeps and trucks in the military, "took me under his wing, just as a friend, essentially. . . . He worked in a lot of different fields, including sheet metal, and I used to go with him to the sheet metal shop after hours."

One of Juris's favorite memories took place in that shop. "It was probably two stories tall inside, and they had a winch for moving huge sheets of metal. It had a little remote so they could raise and lower pieces of metal and move them across the entire shop on these tracks. I think he was sick of me that day, or something, but he pulled over a fifty-five-gallon drum that was empty. He clamped onto the edge of it, put me in the drum, and handed me the remote." Juris extends both hands out in front of his chest, like he's holding a Nintendo console. "So, I'm this kid, cruising myself up probably fifty feet or more, and riding across the shop while he's down there welding. That amused me for hours I'm sure. It's a vivid memory of mine."

There was something more powerful happening in that metal shop, where a kid was whizzing himself around in a fifty-five-gallon drum. "That's where it started for me. It was this real personal experience, where I loved being around someone I admired who worked on these vehicles, and he would include me at ten years old."

Years later, he and his family lacked the money to pay college tuition, so Juris decided to enlist in the air force—primarily "to get the GI Bill." He tested well in mechanics, and there was plenty of work in maintenance, so he became an aerospace maintenance technician.

He worked in that position for three years, ultimately becoming a crew chief. It appealed to him not only because it was hands-on but also because it gave his mind space to wander. "My particular training was to look over aerial refueling tankers, and see if they're ready to fly. And if it has any problems, find someone to fix the problem. . . . I just wanted to learn as much as I could about airplanes and how to get my hands on as many different systems as possible. Because I thought that was fun and cool."

After a couple of years, Juris found himself getting bored. He "had the ambition for something more intellectually stimulating," so he left the air force for college a few days after he became eligible for the GI Bill. He eventually earned a PhD in anthropology from Temple University.

Juris now teaches at Temple and lives in a "big old farmhouse" outside Philadelphia, which gives him a lot of space for maintenance hobby projects. "I have the motorcycles and my wife's car, my car, and a truck. They're all crappy but I keep them running. I enjoy spending time doing that. I think maintaining machines—cars, bikes, and airplanes—is something that is very personal. I gravitate toward it. I feel comfortable there. I'll take on large projects that I've never done before and just teach myself how to do it along the way." In other words, Juris still feels the passion and sense of belonging that he first experienced as a kid surfing a fifty-five-gallon drum through the rafters of a metal shop. His life's journey, like that of many other maintainers, is a story of someone who seeks to align gainful employment with a profound drive for finding joy in his work.

These days, Juris's writing and research applies the tools of anthropology to the environment that gives him the most joy: motorcycle repair shops. In this niche—where he can work on vehicles, teach students, enjoy time with his family, and pursue his scholarship—Juris has found his own way to avoid burnout, keep his passion alive, and make a reasonable living. But to make maintenance work more widely viable, it is not enough to hope that individuals like Juris will simply find their way. Employers need to stop trying to grind every last bit of productivity out of workers; and workers, in turn, need to do more to look out for one another, and avoid "eating their young" (to use the term that troubles the nursing profession). They also need to compensate maintainers at a more appropriate level, one that reflects the value they bring to the organization. Everyone can help simply by appreciating maintenance work of all kinds and acknowledging that it is difficult (and sometimes dirty) work that, if nothing else, provides comfort for the rest of us.

MOWER THOUGHTS

Bill Parslow lit up a smoke and looked out at the lake. "That book you're writing. You gotta talk about the mower thoughts." Bill is a mechanic in the town of Arietta, in New York's Adirondack Mountains. (He's also one of Andy's wife's cousins.) Bill is in charge of maintaining and fixing the trucks, plows, chain saws, and dozens of other pieces of equipment owned by the town. He welded and assembled the bear-proof lids on the town's dumpsters. He orders parts, cleans the shop, keeps everything oiled and lubed, and fixes the axels, bearings, brakes, and whatever else goes out in the morning in good working order but comes back less than 100 percent at the end of the day. When he's not at work he modifies old snowmobiles into racing sleds in the family garage, and keeps up with the fences, firewood, driveways, tractors, and barns on his farm. And then there are the side jobs: troubleshooting a neighbor's generator, changing out the brakes on his friends' trucks, milling parts for his ice shanty, and cleaning the carbs on boat motors. Bill is a maintainer.

"You gotta talk about the mower thoughts," he repeats. One of Bill's side jobs is to mow a few lawns, including at his sister's house on the shores of Piseco Lake. Over the summer months, Bill and the rest of the staff at the town barn work ten-hour days, Monday through Thursday, which leaves Friday open for other things—like mowing. "It's the best time to think," he says, "because you're out here for a while, but it's not like you need to focus on mowing every second that you're doing it."

In other words, mowing, like other routine activities, creates space for your mind to wander. There's a meditative quality to it. Andy thought about the hours he spends on his own yard and keeping his driveway clear when winter storms hit. "It's the same as snowblower thoughts, right?" Andy chimed in. "Sure," Bill shot back. "But it's warmer."

Although there's a profoundly individualistic appeal to mowing and snow-blowing, the communal aspects are just as striking. If you go outside on the morning after a snowstorm, the echoes of shovels and snowblowers seem like they're calling out to each other in solidarity. The same thing happens as evening approaches on a clear summer day, when one lawn mower starts and is soon joined by others. These are technologies that both facilitate introspection and give individuals the power to maintain their corners of the world.

CHAPTER ELEVEN

Caring for Our Homes, Our Stuff, and One Another

On the main drag in Christiansburg, Virginia, there's a shabby, partly vacant strip mall. It has a pawnshop, a Rent-A-Center, a Mexican restaurant advertising $0.99 Taco Tuesdays, and the run-down, empty husk of a former toy store. The parking lot sports a hot dog joint. Deteriorating strip malls are a common fixture of the American landscape, but we bring up this particular one because it's home to the local Habitat for Humanity's ReStore, a warehouse and resale shop that sells home fixtures.

A few weekends a year, Habitat holds what is known as "repair café," with volunteers placed at stations throughout the shop's space. Community members trickle in, carrying broken things—a lawn mower that won't start, pants that need a button replaced—and a staff of volunteers works with attendees to fix them. It's a collaborative effort, one that aims not only to bring the community together but also to encourage self-reliance and environmental sustainability. "We've grown into a throw away culture," Shelley Fortier, the Christiansburg Habitat director, told one reporter. "Our mission is to keep things out of the landfill. Repair, repurpose, revitalize things."[1]

At one of the recent cafés, an elderly resident, Helen Capobianco, brought in a sewing machine that hadn't worked for four years.[2] George Waskowicz, a senior in mechanical engineering at Virginia Tech, opened the machine up and discovered that the problem was a broken cam, which could be easily ordered and replaced. One of us authors brought a set of dull cooking knives and walked out with blades that could slice paper, keener than when they were new.

When we talked with volunteers at the repair café, they often told us that they'd learned the skills they were sharing from parents or other loved ones, and they wanted to pass them down. There was a time when most people—and nearly all women—knew how to sew a button on. Many others knew how to sharpen their own knives, or lived in neighborhoods where knife sharpeners would make regular rounds. It's for this reason that Fortier insisted that the repair café be focused on sharing and teaching skills rather than just having volunteers fix things for attendees. Give a man a fish and he'll eat for a day . . .

One day, a local resident, Judy Ruggles, brought a toaster that no longer worked. While she was standing in line, she overheard repair café organizers Ellen Stewart and Dan Crowder talking about how they also wanted to create a tool library, which would lend out tools to folks who needed them. As Shelley Fortier put it to a journalist, "So many people have deferred maintenance. Often it's because they don't have the tool and may not be able to afford it."[3] Struck by the conversation, Ruggles realized that she'd found the answer both to Stewart's and Crowder's wishes and to the question of what to do with her late husband John's hammers, drills, and chisels.

John had constantly helped people maintain and repair their stuff, something he'd learned how to do growing up on a farm in Kansas. "That's how he met people, by helping them to fix something," Judy later recalled. "That was how he expressed love." Before he died, John, a Vietnam veteran, envisioned a program through which injured veterans—who could no longer do repair and maintenance work themselves but had the knowledge of how it should be done—would act as advisers to people who lacked such know-how. Thinking about the tool library, Judy Ruggles saw a chance for her husband's dreams to come true. The tool library began a soft rollout in the fall of 2019.

While repair cafés, fix-it clinics, and similar events are still relatively rare in the United States, they are part of a growing movement that seeks to help people maintain and repair their own things—or at least be able to take them to local repair shops. These movements bring up a broader question of what we can do to make our world more maintainable, more caring, and thus more sustainable. What would it be like to live in a more caring world? There are many ways to answer that question: We can make improvements as individuals and members of households; respond collectively as communities; and effect change through public policy at the local, state, and federal levels.

In this chapter, we will talk about what individuals can do, but we place the most hope in communal solutions. When it comes to caring for others and maintaining things, too often the feeling is that *we are on our own*. We are more alienated from the things around us today than we need to be. Computers and electronics, for instance, are repairable, and people have been fixing them since they were first introduced. When we share skills and teach one another how to take control of our busted gadgets, it reminds us that we are not alone.

We are not nostalgists. We are grateful for the benefits of modern technology, including the division of labor and expertise that goes along with it. There's a lot to be said for complexity. Similarly, we are grateful for professional caregivers and the systems, like insurance, that give us access to them—although access is profoundly limited by financial inequality. At the same time, there are some unnecessary downsides to our current system that could be better addressed through changes in law and policy. Indeed, the so-called right-to-repair movement highlights how changing these rules opens up new spaces for local businesses as much as it allows individuals to repair things themselves.

When we survey the challenges in maintenance and care that people face in their domestic, or private, lives, it is easy to become pessimistic. But we believe such pessimism is unnecessary. There is reason to hope, and a more hopeful future begins with questioning our desires, especially our hankering for *more,* and then thinking through the—sometimes unhealthy—standards we hold ourselves and others to.

RESISTING THE GROWTH MINDSET AT HOME

Most of us begin seriously considering maintenance in medias res—at the moment when we find ourselves burdened with the costs and work of maintenance we did not expect. A friend of ours has joked that coming to grips with maintenance is a lot like Buddhism—the path begins with suffering. As we've seen throughout this book, one of the things that most contributes to our sense of burden is a growth mindset—as we build and purchase more and more, we bury ourselves under things that require our attention and care.

American homes have been growing, but that doesn't mean yours needs to. When you consider maintenance and rehabilitation costs, how much house do you really need? Hobbies can also burden us. Restoring that vintage sports car might sound like fun, but do you *really* have the time and money to put into it? Or is it going to slowly turn into a rust heap in your backyard?

Take maintenance into consideration when you purchase objects. Look into things you want to buy—from air conditioners to fridges to water heaters—and see if they have known maintenance problems. The repair website iFixit, for instance, takes electronics apart and gives them repairability scores. It can help to remember that the things you own also own you.

If you find yourself in too much of a maintenance or repair bind, you may need to reconsider some previous decisions. Whether that means moving to a smaller home or getting rid of some of the objects that burden your pocketbook, many people find liberation in reducing. Or as Henry David Thoreau once put it, "Simplify, simplify." In *The Grace of Enough,* Haley Stewart, who is Catholic, describes how her family abandoned life in a Florida suburb, where they were unhappy and stressed, for a much simpler but more rewarding one on a small Texas farm with a composting toilet. "We're told that happiness is within our grasp if we can only buy enough, acquire enough, achieve enough. Yet in our pursuit of more, we find only a larger and larger void needing to be filled," Stewart writes.[4] It results in higher maintenance and repair bills as well.

Of course, this caution against materialism is itself a great American tradition that goes back to the Puritans, and we should be careful not to go overboard. Still, there's evidence of a broader shifting awareness around these issues. A few years ago, the tiny-house craze took television by storm, leading to about ten shows on the topic. People enjoy watching and fantasizing about the tiny-house life, even if most people are not adopting it. Only time will tell whether trends like this are part of something larger, or simply another cycle when Americans question consumerism and glorify simplicity (the "back to the land movement" of the 1970s comes to mind).

The study of elderly residents conducted by an aging-in-place taskforce in Blacksburg, Virginia, that was discussed in chapter 7 shows how powerful the habit of putting off maintenance and repair can be, and the devastating consequences that follow. We must face this fact, grapple with the realities we've been ignoring, and plan systematically to take better care. But first we need to think through our ideals and the ends we are trying to achieve.

REJECTING OPTIMIZED LIVING IN THE NAME OF HUMANITY

One of the biggest barriers to creating healthy maintenance and care practices at home is that we've been encouraged to adopt unhealthy ideals of efficiency, optimization, and, ultimately, *perfection* in our private lives. Time and time again, people tell us that they feel completely overwhelmed by maintenance work and the benchmarks against which they measure themselves. They become distraught and near-catatonic—they cannot get anything done.

The historian Ruth Schwartz Cowan, whom we met in chapter 3, encountered these unhealthy standards while writing *More Work for Mother*. One of Cowan's core findings was that supposed labor-saving technologies for the home—like washing machines and vacuums—ironically created more work for mother because cleanliness standards rose with the ability to keep things cleaner. If you have a vacuum, after all, why isn't the floor spotless?

In the postscript to *More Work for Mother,* Cowan notes that many people had asked if writing a book on the history of housework had come to influence her life as a housewife. The answer was yes, but not as quickly as you might expect. A few years after Cowan began research for the book, she observed herself following "the senseless tyranny of spotless shirts and immaculate floors" when her daughter dribbled egg onto her shirt at breakfast.[5] Not wanting her to go through the day with a stained shirt, Cowan made her daughter take it off so she could put it in the laundry, even though an internal voice had begun to question her: " 'What are you doing that for?' I said to myself. 'You know perfectly well that it's the soap manufacturers, and no one else, who foster such absurd notions of cleanliness, and that they do it so as to be able to sell more soap. You, of all people, should not be taken in by such foolishness. Take the shirt out of the laundry basket.' "[6] But Cowan didn't listen to that critical voice. The shirt stayed in the basket.

A few years later, Cowan faced this dilemma again when one of her children got chocolate on her shirt. The internal voice rose again, pointing out that the *child did not care about the stain;* rather it was Cowan's own standards that were driving this quest for gleaming spotlessness. But Cowan still gave in to her superego's urge for perfection—she washed that shirt, too.

Finally, sometime later, Cowan was struck by an illness that kept her on bed rest for half a year. She had always insisted on doing the laundry around the house, but the task now fell to her husband. She badgered him about mixing "darks" with "lights" and "permanent press" with "cotton" until her husband snapped: "He pointed out to me rather sharply (to put it politely) that not a single piece of clothing had been ruined under his tenure." Cowan saw the truth of it. After her health improved, she began sharing laundry duties with her husband. In general, they found that they could ignore manufacturers' guidance about heat settings, not washing colors with lights, and the like, "without causing any damage worse than one slightly pinkened undershirt."[7]

Reflecting on these experiences, Cowan wrote, "Many of the rules that tyrannize housewives are unconscious and therefore potent. . . . By exploring their history we can bring these rules into consciousness and thereby dilute their potency. We can then decide whether they are truly useful or merely the product of atavism or of an advertiser's 'hard sell,' whether they are agents of oppression or of liberation." In this way, Cowan invites us to do some soul-searching and discernment. She concludes, "If we can learn to select among the rules only those that make sense for us in the present, we can begin to control household technology instead of letting it control us. And only then is it likely that the true potential of that technology—less work for mother—will be fulfilled."[8]

If anything, the push for optimization and efficiency has increased since Cowan published her book in 1983. The Internet has become home to subcultures of "lifehackers" questing after ever-increasing productivity. Members of the "quantified self" movement use smartwatches, sensors, apps, spreadsheets, even blood tests to track their vital signs, sleep, moods, diet, finances, work output, and bowel movements. Many of these apps and methods focus on forms of self-maintenance, but what stands out about them is their obsessiveness and their focus on peak efficiency and productivity. The blogging platform Medium is festooned with posts like "42 Productivity Habits to 10x Your Workflow." Makers of appliances, electronics, and gadgets produce an unending flow of thingamajigs—from robot vacuums to self-cleaning toilets—that could be featured in an updated edition of Cowan's book.

But criticisms of this drive to perfection and calls that echo Cowan's plea for sanity are becoming more frequent. In an op-ed titled "We Don't Need to Be Saved from Making Smoothies," the cookbook author and Epicurious editor David Tamarkin lays into what he calls the "Prepared Food Industrial Complex," companies that cast making food from scratch as a waste of time and hawk pre- or partially made meals, all in the name of increased efficiency. Similarly, in *Counterproductive: Time Management in the Knowledge Economy,* Melissa Gregg, a senior principal engineer of client architecture and innovation at Intel, argues that productivity in some ways became a religion during the twentieth century, replacing traditional faiths that aimed at earthly purity and salvation.

In this way, getting "better" at maintenance and care must begin by questioning our *ideals,* asking where they came from and whether they are actually helping us and our loved ones live better lives. We must ask ourselves, When are our things "maintained enough"? When are our loved ones "cared for enough"? Bob Peek, the home inspector we met in chapter 7, said that oftentimes people do not do tasks as simple as changing the filters on their HVAC systems. He can tell because the filter housings are covered with dust, showing they have not been removed for a *long* time. In lots of cases, we don't need to get fancy; we can just start with the basics.

In *The Gifts of Imperfection,* the writer and researcher Brené Brown encourages us to give up comparing ourselves to others—whether it's the perfectly clean and highly designed interiors shown in magazine ads, or the superman who knows the inside and outside of every electronic device and machine he owns, or the overachieving mother who is both a corporate executive and a flawless soccer mom who throws epic birthday parties forever recorded in neighborhood histories. We can always improve things, but we have to begin by embracing our actual lived reality, including much of its inherent crappiness. Anything else is a recipe for stress and insanity.

This process of questioning and reevaluation is something that should not be done alone but *together:* unrealistic, unhealthy, and oppressive ideals of perfection are something we impose on others—including our partners and children—as much as on ourselves. As we have seen in chapter 7, when it comes to domestic maintenance and care, many of these ideals are divided unfairly along gender lines, with the preponderance falling on women's shoulders. If we are to live in a more just society, recognizing the gifts of imperfection and the reality of human limits must be a collective act.

As we've worked to build The Maintainers—much of which is run on volunteer labor—we've increasingly come to think about how we can apply our concepts to our own lives, how we can create a culture of maintenance and care. One of our mottoes is "Make Sure to Maintain Thyself." Our cultural obsession with innovation and growth has not only led us to neglect maintenance of our infrastructure, organizations, and homes, but also encouraged us to be "hustlers" in our work and private lives so that we run ourselves unhealthfully ragged. As a small act of resistance, we've created laptop stickers with the phrase "Maintain Thyself" printed on them. And our friend, the feminist designer and activist Juliana Castro, created Maintain Thyself cards, modeled on the kinds of loyalty cards you see at coffee shops. Only in this case you mark off a square every time you say no to a request. When you say no to ten things, you give yourself a treat or gift. The cards are playful and meant to make people laugh, but they have a serious goal—to incentivize people to care for themselves, realize their limits, and think and rethink the obligations they take on. If we all—including you—are going to be serious about creating a better maintained and more caring world, our thinking has to be more than abstract, it has to begin where we actually sit, live, and work in the world.

THE GOAL IS TO THRIVE, NOT TO BE PERFECT

As we have seen throughout this book, many factors conspire to make us put off maintenance. At least some of these factors are psychological. Studies suggest that hyperbolic discounting—the tendency to choose a small reward now (e.g., going down a YouTube rabbit hole) over a larger reward later (e.g., having an enjoyable, well-maintained house)—is at least partly a matter of cultural rearing. The bad news for Americans, when compared to other nations like South Korea, is that *we do have a tendency to discount* the future in favor of short-term pleasures. Moreover, while some people take great pleasure in maintenance work, many others find it dreary and boring and will do *almost anything* to procrastinate and put it off. The act of deferring maintenance is overdetermined.

For all these reasons, getting better at maintenance and care must begin by facing up to our propensity to put them off. We can recycle an old chestnut from the twelve-step world: "The first step is admitting you have a problem." As the environmental author and yoga instructor Eileen Crist Patzig once put it to one of us, "People tell me all the time that they don't have time to build practices like yoga or other forms of exercise into their schedules, but the truth is they don't have time not to."

The best way to do this is to create practices that lead us to be mindful of maintenance *systematically*. By "systematic," we mean in a planned and methodical way. As we've worked on these topics over the last six years, we've witnessed many diverse methods individuals have developed to keep on top of maintenance. One of us has an aunt who keeps a notebook in her car that lists all of the major automotive maintenance tasks, when they were last done, and when they next need doing. She checks this notebook regularly, so she'll know when to schedule repairs. Others have told us about keeping spreadsheets of home maintenance tasks, or files that list the contractors they have hired to do work, including what was done and how much it cost. We have also experimented with home maintenance software and apps, like HomeZada, HomeBinder, Home Management Wolf, The Complete Home Journal, and MyLifeOrganized. You can also find home maintenance checklists online. Additionally, apps like Centriq help you keep track of and repair home appliances, including warning you about recalls.

These kinds of planning systems have their limits. People often grow immune and less responsive to calendar reminders over time—at least that's how it works in our experience. If some digital system has been prodding you to go to the gym and yet you have not worked out for the thirtieth straight day, something isn't working. In such cases, it's often best to start over. Redesign your plans; ask if they are realistic; make them simpler. Attending regularly to such systems is the key to making them successful.

Moreover, even with good preventive maintenance, repair-based, reactive maintenance problems will pop up. Shit happens, often at the most inopportune moments. Murphy's Law exists for a reason. You can't "plan" for such happenings, other than by setting aside money for when they occur, which is why the two of us have created maintenance savings accounts specifically for this purpose. According to some surveys, the average American household spends $408 per car on annual maintenance—that's about $35 a month, without factoring in car payments, gas, or any other auto-related costs.[9] One of us lives in a log cabin home, and by some reckonings, wood houses require as much as $3,000 of maintenance every two years, or $125 a month. That's not nothing. And when you start to stack these costs together, they add up quickly. Of course, the terrifying thing here is that about 80 percent of American households live paycheck to paycheck, with few of them having explicitly budgeted for maintenance.[10] (About 60 percent of households do not budget at all.)[11] If you're a household struggling to make ends meet, an expensive repair job can send your finances into a tailspin. It's a bleak reality.

Still, even in less than ideal financial circumstances, planning helps. In the course of writing this book, we both adopted the practice of holding regular family meetings to organize things like our budget, exercise regimens, supplements to the children's education, chores, and meals. We use the meeting to plan care and maintenance, and to include our kids in the work that goes into running a home. (Because our children are still young, the parents do most of the meeting alone, but the kids are brought in for meal planning. We have found that including them in food choices reduces their griping later on, though not universally. Kids will be kids.)

As we have come to better understand the homes we recently bought and moved into, we're getting a grasp on how much maintenance was deferred by the previous owners. Family meetings have helped us keep this reality front and center in our minds. Is spending $2,000 on a vacation really a good idea when it will cost $15,000 or more to eventually replace a heavily used deck before it becomes unsafe? Does the family need another tablet computer, or should the savings be put toward the $10,000 it will take to clean, stain, and repair the wood exterior of Lee's log cabin (something that has apparently not been done since his home was built in 1983)?

We have adopted the systems we use to run these family meetings from the corporate and nonprofit worlds, including from individuals and groups who have been helping us to develop The Maintainers. This seems like an ironic twist, given that we just warned about the dangers of taking on the unrealistic ideas of efficiency and optimization that are so common in the working world. Are we now suggesting that we "lifehack" domestic maintenance and care? No and yes. No, we do not want to hold our private and family lives up to standards of "peak efficiency" or "return on investment" or whatever other hellish concept is in vogue at a given moment. But, yes, we can draw on budgeting methods, software, apps, and other tools to help us structure, remain aware of, and keep on top of things. Indeed, one of the most important tasks for us is to schedule unstructured time to maintain ourselves—whether that means taking a family hike in the woods, playing in the yard, or pulling out some board games. Structures help us to protect this time together from the encroachments of work, chores, errands, and other tasks, keeping it sacred.

Putting money into maintaining, rehabilitating, and even replacing parts and systems of our houses is immensely rewarding to us. It comes with a feeling, even a thrill, of maturity and responsibility—emotions that are too little appreciated in our culture—knowing that you have taken care of and ensured the longevity of the place where your family spends most of its time and has its most intimate, loving, and affectionate moments.

When we were writing this book, a flyer from Lowe's home improvement store appeared in Andy's mailbox one day. It featured a family playing in a crystal clear pool, with coupons for pool chemicals on the back and a slogan on the front: "Enjoy More. Maintain Less." As the owner of a pool himself, Andy found the (false) dichotomy between enjoyment and maintenance jarring. In Andy's experience, pool maintenance is in fact a joyful task. There's the challenge of balancing the chemicals, the sense of accomplishment that results from cleaning the filter and liner, the smell of the bleach and chlorine, and the fleeting moments outside in the quiet evening air. These and other chores accumulate meaning every time he watches the kids jump in and smile. A well-maintained pool is a place where family and friends can gather, commune, and revel in precious summer days and evenings.

To us, all of this suggests that there may be a gap in popular discourse about the things we own and use. On one side is a kind of consumerism that celebrates the rapid consumption of disposable things. On the other are traditions that criticize "materialism" and assert that objects are not the answer to our soul's desires. What's missing is a kind of positive materialism that recognizes the deep pleasure and meaning that can accompany physical realities. It is not surprising to us that a pool is an object *we enjoy with others,* not an object of solitary consumption like staring into a smartphone screen.

IMPROVING THE LAWS THAT REGULATE MAINTENANCE AND CARE

Many improvements to maintenance and repair can only be achieved collectively. Federal law and policy will be helpful, and in some cases necessary, but most of these refinements need to be addressed at the state and local levels, specifically the latter. Creating online communities—and sharing information through them—is important as well.

After realizing that many people didn't have the skills to repair things and were throwing them out, the engineer Peter Mui founded a group called Fixit Clinic in 2009. Fixit Clinics are a lot like Habitat for Humanity's repair cafés, except Mui's vision focuses much more on teaching people the skills of repair. At some repair cafés, you can bring in something broken and have someone fix it for you. But at Fixit Clinics, the "coaches" work with people to help them perform the fix on their own. "At Fixit Clinics, we ask the participants to materially participate in their own repair," Mui emphasizes.[12] The real aim, he says, is a kind of conversion. Coaches want people to realize it is okay to open things up and see what is wrong. It's about enabling people to troubleshoot and fix things without feeling frightened of them.

Fixit Clinic has a ritual. When someone comes into the clinic, the organizers announce their name, their item, and the item's symptom(s) to everyone gathered. "Hey, everyone, say hi to Ted and his DVD player that skips."[13] They also place similar repair jobs next to each other. What they find is that participants become invested in each other and each other's projects.

Mui hopes that Fixit Clinics do more than just embolden people to repair things. He wants it to shape how they act as consumers, too: "[That's] where we're ultimately going with Fixit Clinic: to encourage products designed with maintenance, serviceability, and repairability in mind. As consumers, we're going to have to start demanding those things; at Fixit Clinic, we trust that improvements . . . will come through a broader understanding and dialogue around how things are made now."

Repair cafés, fix-it clinics, and other such gatherings focus on local, in-person interactions. We think that's a good thing. And while not everyone has access to such events, there are many tools for distributing knowledge about repair. Boosters of the Internet in the 1990s, such as Grateful Dead lyricist and Electronic Frontier Foundation cofounder John Perry Barlow, promised that high-quality information would be diffused near and far. It's easy to make fun of that optimism now, in an era of fake news, hacked elections, partisan rancor, hate speech, and online bullying (not to mention the predominance of cat videos). The Internet is home to countless expressions of stupidity, irrationality, and nastiness, but the predictions of Barlow and others hold true in some ways for websites devoted to repair.

Sites like iFixit and YouTube have become massive libraries of how-to and DIY materials. An Uber driver regaled one of us with the story about how he now repaired and maintained every part of his car on his own, even though a few years earlier he had known almost nothing about cars. He learned it all from watching YouTube videos. Today, iFixit hosts nearly fifty-five thousand repair guides on its site and partners with more than eighty universities around the United States to teach students about technical writing and the importance of repair. As a result, more than nineteen thousand students have created the majority of the repair guides—and people are using them. In 2018, iFixit saw 120 million unique visitors to its website—with 7.8 million unique visitors from California alone, or nearly 20 percent of the state's population.

Institutions also use digital tools to organize and manage tool libraries and timebanks. Tool libraries, from which individuals can check out a tool as one checks out a book from traditional libraries, contribute to thrift and environmental sustainability beyond the important task of keeping things from being thrown away or degrading. Timebanks are another way for people to share skills and maintain things or get them repaired. Here's how it works: People volunteer their time and skills—sewing, tutoring, pet or child care, yoga or meditation sessions, plumbing, and so on—to fulfill other people's needs. When they complete the work, they bank time credits that they can use in turn to hire someone else to do something for them. The idea is to open up skills-sharing economies within communities.

Civic groups in Virginia are experimenting with collective ways to help families with so-called critical home repair needs: broken things that threaten to force them from their homes.

The Floyd Initiative for Safe Housing (FISH), run by Habitat for Humanity, works to ensure that families can stay in their homes; its name alone suggests that *unsafe* housing is a threat to communities. FISH leader Susan Icove—a civic-minded ceramicist and potter whose current work focuses on candlestick holders, lamps, and other lighting—described visiting a mobile home where residents were using the oven as their heater, and the only electricity in the house came from an extension cord run from next door. In another mobile home, residents in their seventies had gone without running water for six months. Habitat director Shelley Fortier notes that 22 percent of the housing in Floyd County consists of mobile homes that are often located in trailer parks. Made from cheap materials, mobile homes degrade faster than other kinds of housing. This creates real problems; for example, floors and ceilings sometimes collapse without much warning. FISH aims to do eight to twelve repair jobs a year—to be eligible, the job cannot cost more than $2,000 and involve no more than two days of work. But the group is sometimes overwhelmed with requests, receiving one or two referrals each week.

FISH relies completely on local donations. People who work on critical home repair have repeatedly complained to us that state-level programs require repair jobs to bring homes "up to code" (electrical, fire, etc.). But while the intention of such requirements is good, bringing many of these dwellings up to code is *impossible* with the amount of money available. It's one of those instances when perfect is an enemy of good. FISH focuses instead on doing whatever small jobs are needed to keep people safe, dry, and in their homes, when *no other options are available*.

As much as we salute these experiments with community-based ways of improving maintenance and repair, there are barriers to these activities that can only be removed by changing laws. Remember the story of Kyle Wiens, the CEO and editor in chief of iFixit, in chapter 7? When Wiens broke his Apple laptop as an undergrad and posted his own repair guide online, he learned that Apple was using copyright law to keep its repair manuals off the Internet. It had a bunch of other techniques for shutting down independent repair work, too.

Others were bumping into similar frustrations. For many years, Gay Gordon-Byrne ran an independent consulting company focused on buying, selling, and leasing computer hardware. She became personally outraged watching companies buy technologies and then sign end-user agreements that prohibited them from repairing the technologies themselves or hiring independent repairpersons.

For the large Fortune 500 companies Gordon-Byrne often worked with, paying more for repair was no big deal, but she saw for herself the negative way in which these repair restrictions affected small businesses and the self-employed. Repair restrictions spread throughout the early years of the 2000s, but in Gordon-Byrne's estimation, the trend took off around 2010: "We woke up one day and said, 'Holy Cow.' "

Gordon-Byrne, Wiens, and other members of iFixit partnered in 2013 with a handful of other organizations, such as the Electronic Frontier Foundation and the Service Industry Association, to form the Digital Right to Repair Coalition, later simplified to the Repair Association. The focus of the Repair Association is to change state laws to require manufacturers to make information and parts available to consumers and repairpersons. In the last few years, more than twenty states have introduced or debated right-to-repair bills (though none have yet become law). Gordon-Byrne says that the right to repair requires a "five-legged stool" approach. To do a repair, you or someone you hire needs (1) a manual, (2) parts, (3) tools, especially given that companies use odd-shaped, specialized parts to limit access, (4) the ability to read and understand computerized diagnostics, including knowledge of what the strange error codes that show up on our gadgets mean, and (5) access to firmware (low-level software used to control hardware) and passwords that manufacturers use to lock down repair. Without these five elements, it's extremely difficult for owners to fix their own property, and the aftermarket cannot thrive.

Companies defend repair restrictions in various ways, including by playing up fears around cybersecurity and product safety. Some of these claims appear to be specious or overblown. Right-to-repair advocates have yet to see a documented case of someone injuring themselves while changing a cellphone battery, for instance. Lobbyists for Apple also told Nebraska lawmakers that if they passed repair legislation they'd be making the state a "Mecca" for hackers.[14]

Often the real reason for industry resistance is simpler. It's about money. The Morningstar analyst Scott Pope estimates that repair work at John Deere dealerships has profit margins that are five times higher than that of selling new equipment. Apple can charge as much as $1,000 more for a repair than a local repair shop will charge you.[15]

Cellphone makers, appliance manufacturers, and many other firms that use repair restrictions don't fit the Federal Trade Commission's antitrust definition of monopoly, which requires a producer to control 75 percent or more of a market. But Richard John, a professor at Columbia Journalism School who is currently writing a book on the history of antimonopoly crusades, points out that "monopoly" used to have a broader definition. Monopoly was "any kind of market power that was conceived as unfair, any power that gave an institution an unfair advantage," John explained to us, a definition that certainly fits how right-to-repair advocates describe repair restrictions. "Antimonopoly is a Main Street value," he said. "Historically, it was primarily backed by Main Street Republicans."

While the rights of consumers and the plight of individual farmers get the lion's share of media coverage around the right-to-repair cause, advocates often focus on the broader business impacts. Kevin Purdy of iFixit recently published "Right to Repair Is a Free Market Issue," which examines how anticompetitive repair restrictions shut down independent repair shops.

Kyle Wiens, the CEO of iFixit, points out that companies like Apple have not focused on building repair businesses because the profit margins are so low. But small businesses will go after these market niches, Wiens explained to us, because "they can be tackled with lower overhead than the big manufacturing has. Those small businesses are providing an additional service to the market; they're providing liquidity; they're providing consumer value; they're creating local jobs; they're creating more self-reliance and busting up monopolies a little bit." Right-to-repair advocates estimate that there would be hundreds of thousands more independent repair shops if repair restrictions were lifted.

Wiens, who learned how to repair things from a beloved grandfather, emphasizes that repair comes with a sense of pride. He points to Matthew Crawford's book *Shop Class as Soulcraft*, in which the author describes how he left behind work in universities and think tanks to become a motorcycle mechanic. "Crawford talks about [how] there's a pride and a satisfaction that he has from his community respecting his work that he didn't have as a think-tanker," Wiens told us. That sense of independence and pride can be hard to quantify, but it's still an important reason to fight for our right to repair.

There are other kinds of legislation and policy making that can improve maintenance and repair. Sweden has introduced tax breaks to encourage individuals to repair goods rather than throw them away.[16] We can envision tax-sheltered accounts for maintenance and repair similar to the tax-free education savings accounts that already exist. Moreover, in some areas, legal changes will be necessary if we want to live in a more maintainable world. Martine Postma, the founder of the Repair Café International Foundation, has lobbied for legal and policy changes that would move us toward a "circular economy"—increasing the recycling and reuse of materials and vastly decreasing our waste stream, with the ideal being near zero waste. Given how many manufactured objects—including most of the plastic articles in our lives—are not recyclable, a true circular economy would require a profound shift in our culture, likely including tough regulations to force manufacturers to make their goods thoroughly repairable and recyclable.

What's most striking about many of the concepts discussed in this chapter—from repair cafés to right-to-repair laws—and previous chapters is how they appeal to different kinds of people, including both conservatives and liberals. We are hopeful about the future of maintenance and repair because new conversations are opening up about how we can improve the upkeep of public infrastructure, better the lives and jobs of maintainers, and address some of the toughest questions in our culture, including how bad maintenance unduly burdens poor and marginalized communities. It is to these new conversations, and the invitation that comes with them, that we now turn.

EPILOGUE

From Conversation to Action

We wrote this book to raise awareness about maintenance and elevate the status of those who perform care and repair work. The task ahead of us all is to cultivate richer and more productive conversations, and to use those conversations as fuel for collective action. Conversation is important—and we see great potential for more productive conversations about maintenance, as we will explain in a minute—but conversation alone is insufficient. Likewise, action without conversation is also insufficient. As we saw in part 3, today there are plenty of people taking significant actions to advance the cause of maintenance—but these actions are often sporadic, isolated, and lacking in the scale and intensity to add up to a powerful social movement.

We do not believe that our society lacks the financial resources or technical expertise to become better maintained and more caring. Sometimes maintenance is too expensive, or it's difficult to find the right people for the job. But those are problems that can be resolved. We live in an era of great wealth, and our educational institutions, imperfect as they are, remain powerful engines for nurturing resourceful and talented people.

We have seen firsthand the creativity that happens when people push past these common narratives and envision a different way of maintaining our world. We (Lee and Andy) are two of the three codirectors of The Maintainers, a global interdisciplinary and interprofessional community that examines maintenance, repair, infrastructure, and the ordinary work that keeps our world going. Together with our third codirector, Jessica Meyerson, we have organized a variety of activities that include conferences, video discussion groups and seminars, email and social media chatter, and focused, in-person convenings of experts in specific fields like digital archives or workforce development. From these activities we have contributed to coalitions around policy issues, such as the right-to-repair campaign, and we have secured grant funding for projects to build advocacy tool kits for librarians and archivists.

Below, we'll recap some of the actions we see as essential for a better-maintained world—many of them have already been described in the preceding chapters. But first we'd like to focus on some of the conversations we've been joining and facilitating through our work with The Maintainers, and why we're so convinced that these communities have a lot of potential as foundations for future action. The most enjoyable part about them has been the mix of people who share our core concern—that the Innovation Delusion is damaging our society—as well as our core hope that a revaluation of maintenance will lead to a healthier society. We'd like to highlight a few of those people, from vastly different walks of life, and with varied professional and political commitments, whose passions converge on their work to make a better and more cared-for world.

BUILDING A MOVEMENT

One of the frequent contributors to The Maintainers email list is Camille Acey, whom we met in chapter 10. At her day job, Camille works at a software start-up, where she maintains good relationships between her company and its customers. But her ethos of maintenance and care runs much deeper than that. Camille's parents raised her with a "fight the power" mentality, and she has been a community activist in New York City for more than twenty years. Her father, who grew up in Newark, New Jersey, took part in the Black Arts Movement that has been characterized as the "aesthetic and spiritual sister of the Black Power concept."[1] Camille's mother, a nurse, was active in her union. After the 2016 election, Camille started quoting the African American feminist writer Audre Lorde in her email signature: "Caring for myself is not self-indulgence, it is self-preservation, and that is an act of political warfare."[2] Camille emphasizes that maintenance is most effective when it is infused with a spirit of care.

Gracy Olmstead is on another part of the political spectrum, but she shares Camille's concern about care for self and community. Olmstead, a writer from Idaho, comes from several generations of farmers and grew up in a small town that has experienced significant decline in recent decades. Her work focuses on the societal value of farms and visions of *regenerative agriculture,* practices that rebuild soil matter, restore biodiversity, and fight environmental harms like climate change.[3] Central to this vision is a rejection of what conservative and Catholic thinkers—including Pope Francis in his encyclical *Laudato si'*—have come to call throwaway culture. As she put it to us, "We throw away many wonderful, beautiful things, and we treat both our planet and ourselves very badly when we tap into this very consumptive culture."

Like many Americans, Olmstead's beliefs do not fit easily into our current political landscape. As a Christian raised in an evangelical household, she cares deeply about pro-life politics and identifies as a conservative, but she also finds that other central values, such as a focus on social justice, community wholeness, and environmental protection, do not fit within the current Republican Party. She is a political nomad. Olmstead describes having been influenced by her husband, who does maintenance work for a living as an avionics technician. "He repairs everything in our home, from our dishwasher to our laundry machine to figuring out how to install plumbing so that we don't have to pay someone else to do it to taking apart my computers when I spill random beverages on them."

Olmstead hasn't attended any Maintainers meetings, but she contributed to the conversation in 2016 when she wrote about our essay "Hail the Maintainers" for *The American Conservative:* "Many people associate the word 'innovation' with Republican sentiment, because the party prizes capitalism, free markets, and entrepreneurship. But to be a conservative is also, importantly, to desire to *conserve* things. To appreciate the quotidian labor that keeps our world going—and to join the maintainers in tending our little square of earth, keeping the weeds out of our gardens with the same diligence and zeal with which we wash our faces."[4]

Appreciation for this kind of labor is a value shared by the computer programmer Björn Westergard. He lives and works in Washington, D.C., where he is currently a senior software engineer at National Public Radio. In high school, he had an eye-opening experience while interning at a defense contractor. "Up until that point I had been led to believe that you should pursue a technical occupation because it contributed to the common welfare in some sense," he later told a reporter.[5] He became interested in labor movements and socialism. In early 2018, Westergard and his fellow computer workers informed their employer, Lanetix, a San Francisco–based transport and logistics company, that they intended to unionize. Lanetix quickly fired all fifteen of them, and the workers filed a petition against the company with the National Labor Relations Board. Lanetix settled in November 2018, paying the workers a total of $775,000. Westergard's experience meshes with the research findings we cited in chapter 8: Most IT workers do maintenance-type labor.

Westergard moderates a Facebook community called the Relaxed Marxist Discussion Group. The group's profile photo shows Fred Rogers of *Mister Rogers' Neighborhood* in one of his trademark red sweaters—only in this case, the sweater features the hammer and sickle symbol. Marxists are famously sectarian and fractious, and

Westergard does a fair amount of group moderating work to maintain a *relaxed* environment, including pointing members to other Marxist discussion groups where they are permitted to fight. Westergard attended the first two Maintainers conferences and has participated in a range of discussions within the network, and he continues to focus on workers' rights and creating a more just society.

A fourth person we'd like to highlight, Chuck Marohn, the founder of Strong Towns, who is profiled in chapter 4 of this book, was a keynote speaker for our Maintainers III conference in Washington, D.C., in October 2019. Marohn grew up and lives in the small town of Brainerd, Minnesota. A conservative Catholic with libertarian leanings, he overthrew many of the assumptions he'd learned from being an engineer and a city planner. Marohn's insistence that change starts with small communities resonates with us: There are plenty of things that the federal government can do to help, but we find it unwise to expect some well-considered program devised by bureaucrats to properly fund maintenance activities and to respect the dignity of maintenance and repair workers. We believe Camille, Gracy, and Björn would agree.

From the beginning of our adventure with The Maintainers, we realized that discussions around maintenance and repair slipped the surly bonds of traditional left-right politics. Members of The Maintainers network reread both the works of left-leaning thinkers, like Karl Marx and Hannah Arendt, and the works of conservatives, like Edmund Burke and Michael Oakeshott, to see what they had to say about maintenance. Such bipartisan interest is unusual, and perhaps even surprising, in our fractious, divided moment. But we've learned that partisan politics and identities fall to the wayside when people give themselves and one another some space to talk about maintenance and repair. These subjects are more engaging, more urgent, and more promising than the clickbait political issues of the day.

We have also been surprised by how many people have found themes of maintenance and repair in their religious traditions and spiritual lives. At our second Maintainers conference, Varun Adibhatla, a public technologist who is using digital technologies to improve maintenance practices, put up a slide that juxtaposed two images. On one side were three faces: Steve Jobs, Elon Musk, and Archibald "Harry" Tuttle, a character played by Robert De Niro in Terry Gilliam's 1985 movie *Brazil*. Tuttle is a renegade underground repairman who subverts the system by slipping past bureaucracy and giving people needed maintenance. On the other side was the Trimurti, the triple deity central to Hinduism: Brahma the creator, Vishnu the preserver, and Shiva the destroyer. Varun's point was that our current Silicon Valley–centric narrative, based around people like Jobs and Musk, is too focused on creation and change and not enough on preservation. The narrative lacks the balance that Varun finds in the central myths of his Hindu faith.

Others have found ideas of maintenance in the monotheistic traditions of Judaism, Christianity, and Islam. It's something the Presbyterian minister Fred Rogers pointed to when he was asked to make a public service announcement after the September 11 attacks. He looked into the camera and said, "No matter what our particular job, especially in our world today, we all are called to be *tikkun olam,* repairers of creation."

The political and religious salience of maintenance points toward a deeper philosophical basis for a social movement in support of maintenance and maintainers. Bonita Carroll, an Australian ethnographer who studies maintainers in the mining industries, once described maintenance to us as a movement the world has yet to adopt: "I remember when the [government] gave us all the colored bins to separate trash [and recycling]. We had to actively train ourselves to use them. But as time has progressed, it's naturalized, normalized, and is innate. Adopting the maintenance mindset takes our scope those few steps further, promoting an understanding and appreciation of the entire lifespan of objects."

It's important not to be naïve about the political implications of all of this. Progressives and conservatives will fiercely advocate for different solutions even when they both recognize a problem. Take the plight of maintenance workers, who, as we saw in chapter 6, make up the vast majority of the working poor in the United States: Progressives may argue that raising the minimum wage would help the many maintainers and their families who are stuck in poverty, while conservatives believe that mandatory wage hikes are a proven way to punish the poor by eliminating jobs.

Still, we are heartened to see different kinds of people connect over the topic of maintenance in ways that they cannot when discussing hot-button "culture war" topics. And we believe that the different groups and sides can learn from one another. You do not have to share Chuck Marohn's aversion to federal programs and spending in order to benefit from his insights about how local governments should rethink their infrastructure. You do not have to be an environmentalist or an advocate for a "circular economy" to believe it's important to make technologies more maintainable and repairable. At repair cafés and fix-it clinics, where the desire to be together and be polite avoids party politics altogether, the left and the right literally learn from each other—one turn of a screwdriver at a time.

Some people have argued that focusing too much on maintenance would lead to a defense of the status quo. We hope that by now it is clear that our target is the ideology of innovation-speak, not innovation itself. Actual innovation has a complex relationship to maintenance and repair. Furthermore, economic and other forms of inequality have *increased* during the most intensive period of innovation-speak, from the 1970s to the present. The ideology of innovation is no solution to these problems.

CALLS TO ACTION

We're well aware that changes in language and jargon do not by themselves produce measurable changes in material or social conditions. We'd be silly to believe that the end of innovation-speak would mean the problem has been solved.

How do we move from thinking about these issues to action? Our first step was to establish The Maintainers to generate communities, and then to work with Jessica Meyerson to increase The Maintainers' capacity to cultivate those communities and facilitate conversations. What intrigues us most is learning why the subject of maintenance excites various individuals—or at least engages them. We often begin conversations with questions: What do you want to see maintained, and why? Who or what do you maintain? Who or what maintains you? Who do you see as model maintainers in your life?

We promise you that if you join The Maintainers, we will not present you with ready-made, off-the-shelf fixes that will make all of your problems go away. We do not believe such solutions exist—as we pointed out earlier in this book, part of the wreckage of our culture stems from how the consultants of the world promise innovation and growth via one-size-fits-all approaches. Like Chuck Marohn and others, we think that such "systematic" techniques mislead us and blind us to the on-the-ground, concrete realities of our lives. We can, however, learn from one another and share tools that we have found helpful for our particular situation. What you will find in The Maintainers is a community of people working together, wrestling with the question of how to care for people and things better. We are motivated by visions and responses to this question: What would it be like to live in a more caring world?

We've found it most helpful to think in terms of three scales of action: societies, organizations, and individuals. We organized parts 2 and 3 around these three different scales of action: Chapters 4 and 9 focused on the societal level; chapters 5 and 8 on organizations; chapters 6 and 10 on the work that individuals do; and chapters 7 and 11 on the ways that individuals encounter innovation-speak and maintenance in their private lives and homes.

These three different scales of action are evident in two examples that are very important to us, in which maintenance plays a key part. They are, first, injustices around race, poverty, and disability and, second, climate change. We see these as existential matters. Without progress on such issues, there are real reasons to worry about the sustainability of our society as we know it.

Like other forms of wealth, maintenance is unevenly distributed.

Neighborhoods that are poor or segregated by race or ethnicity are typically poorly maintained. Often enough the truly disadvantaged lack access even to mundane modern technological systems—for example, the absence of water management systems in Lowndes County, Alabama, that is responsible for the resurgence of hookworm infections described in chapter 4. But even when they do have access, those systems are badly maintained. Subway stations in poor neighborhoods in New York, Washington, D.C., and other major cities are ugly and run-down. Public housing residents wait days or weeks for broken elevators to be repaired. Students in failing public school systems attend class in buildings with leaking roofs, busted furnaces, and lead paint chips. Poor tenants face landlords who refuse to maintain their buildings, while impoverished homeowners lack the means to stave off decline. Moreover, as disability rights activists will tell you, poorly maintained wheelchair ramps and out-of-service elevators indicate that an organization is not taking disability seriously.

In this way, whatever justice looks like, it will require constant upkeep and care.

The same goes for climate change. Addressing it in a meaningful way will require significant technological change, a much deeper metamorphosis than many people grasp. All of these new technologies would, of course, require maintenance, and the movement for environmental sustainability can succeed only to the extent that it adopts a maintenance mindset. We can easily imagine renewable energy sources funded by the Green New Deal or some other large federal program that then *fails* to maintain them. We hope that representatives in Congress understand this dynamic when they are considering multibillion-dollar infrastructure bills—which are perpetually touted as one of the most likely areas for compromise across our partisan divide.

The existential urgency raised by activists and advocates of the Green New Deal raises the issue of survival itself—not a pleasant topic, of course, but one that all responsible adults have the burden to contemplate. When technologies are harmful, we should stop maintaining them and instead help them die—put coal-burning power plants in hospice, give palliative care to coal mining. We can find creative ways to reuse these systems and their spaces to benefit our lives, from turning industrial sites into historical parks or transforming railroad beds into trails, as New York City did to create its much-celebrated High Line.

We wrestle collectively with difficult questions because they do not have easy answers: How do we deal with and pay for the current large backlog of deferred maintenance? How do we reorient federal infrastructure policy to help localities deal with infrastructure rather than burden them with systems they cannot afford? Should we view access to clean water, functioning electricity, and other basic modern infrastructures as a human right? How do we change our culture to value maintenance and focus on it so that we stop getting into the same messes? How do we ensure that maintainers—the people who keep our society running—are recognized and adequately compensated? How do we help individuals and families with the burdens of domestic maintenance and care that affect so many of us? As we have seen throughout this book, many maintenance problems put the most strain on the poor, the elderly, and the least among us. Ask yourself: How do you want your loved ones to live and die? How do *you* want to live and die? We would be lying if we told you we knew the answers to these fundamental questions. But this book has given us a better sense of where to look for them.

How do you want this to go down? You tell us—we're listening.

Acknowledgments

It's customary for authors to express the futility they feel in composing acknowledgments, given how many people supported the work of writing a book. So you can only imagine how overwhelmed we are feeling as we complete this book about the need to celebrate the maintainers whose labor keeps the forces of entropy at bay. We're delighted to name a few of the people who helped us along the way.

Sam Haselby, senior editor at *Aeon,* was the first person who helped us turn our jokes about maintainers into a serious examination of the subject. Lauren Sharp at Aevitas Creative encouraged us to think about our project as a book, and guided us into the world of commercial publishing. We lost track of how many times we thanked our lucky stars for being able to work with Derek Reed, our gifted editor at Currency. The production team at Random House, particularly the intrepid copyeditor Maureen Clark and the senior production editor Robert Siek, tidied up some of our clunky prose and tenuous claims. We are grateful to them and all of the people in unseen, unsung "maintainers" roles at the press who kept this book rolling along, as they do for the publishing industry. We are likewise fortunate to have the joyful, enthusiastic, ALL CAPS ENERGY presence of our Maintainers codirector, Jessica Meyerson, in our lives. We are so grateful to all the members of The Maintainers network, including the individuals on the mailing list and those who have attended the conferences, for providing a community to develop these ideas. Finally, we interviewed dozens of people for this book, and want to express our appreciation for their time, energy, and candor.

There are too many of our peers, friends, and colleagues to mention here, but we feel special warmth and gratitude toward David C. Brock, Jenni Case, Juliana Castro, Nathanial Comfort, Ruth Cowan, David Edgerton, Brad Fidler, Yulia Frumer, Lou Galambos, Seth Halvorson, John Horgan, Samantha Kleinberg, Scott Knowles, Bill Leslie, Theresa MacPhail, James E. McClellan III, Patrick McCray, Bethany Nowviskie, Bill Parslow, Brad Parslow, Phil Scranton, Brian Shaw, Eric Stotts, Steven Usselman, Heidi Voskuhl, and Ben Waterhouse.

Andy thanks his parents, Larry and Carol Russell, for their patience and support; and his wife and kids, Lesley, Reese, and Calvin, for the joy they bring to the world.

Lee thanks his wife and kids, Abigail, Henrietta, and Alban, and his trusty pooch, Baron, for making his life worth maintaining.

The research for part of this book took place while Lee was a fellow at the wonderful Linda Hall Library, a science and engineering library in Kansas City, Missouri. We are very grateful to Linda Hall for providing research support as well as for sponsoring Maintainers Happy Hours. Extra special thanks to Linda Hall's Ben Gross for his guidance, friendship, and encouragement.

This book is dedicated to all the maintainers who keep the best and most necessary parts of our world going—including the teams of cleaning and maintenance staffs at our respective employers; the plumbers, electricians, roofers, and other skilled tradespeople who maintain and repair our homes; the medical professionals who look after our health; and the employees who maintain the power, sewer, data, and transportation systems that we often take for granted—and to the Society for the History of Technology, our intellectual home where this work began.

Notes

CHAPTER 1: THE PROBLEM WITH INNOVATION

1. Malcolm Gray, "Hidden Threats from Underground," *Maclean's,* September 1, 1986, 83.
2. Henry Blodget, "Mark Zuckerberg on Innovation," *Business Insider,* October 1, 2009, https://www.businessinsider.com/mark-zuckerberg-innovation-2009-10. In full Zuckerberg said, "One of the core values of Facebook is 'Move fast.' And we used to write this down by saying, 'Move fast and break things.' And the idea was, unless you are breaking some stuff you are not moving fast enough." Within ten years, Zuckerberg was apologizing to lawmakers in the United States and Europe on account of Facebook's abuse of customer privacy.
3. Peter Manzo, "Fail Faster, Succeed Sooner," *Stanford Social Innovation Review,* September 23, 2008, https://ssir.org/articles/entry/fail_faster_succeed_sooner.
4. Jonathan M. Ladd, Joshua A. Tucker, and Sean Kates, "2018 American Institutional Confidence Poll," Baker Center for Leadership and Governance, Georgetown University, https://bakercenter.georgetown.edu/aicpoll/.
5. Jim VandeHei, "Bring on a Third-Party Candidate," *Wall Street Journal,* April 25, 2016.
6. https://news.gallup.com/poll/1678/most-admired-man-woman.aspx.
7. Dominic Basulto, "The New #Fail: Fail Fast, Fail Early and Fail

Often," *Washington Post,* May 30, 2012, https://www.washingtonpost.-com/blogs/innovations/post/the-new-fail-fail-fast-fail-early-and-fail-often/2012/05/30/gJQAKA891U_blog.html.

8. Nadeem Muaddi, "Florida University Used Time-Saving Technology to Build Its Collapsed Bridge," CNN, March 16, 2018, https://www.cnn.com/2018/03/15/us/fiu-bridge-collapse-accelerated-bridge-construction/index.html; Alan Gomez, "Miami Bridge Collapsed as Cables Were Being Tightened Following 'Stress Test,'" *USA Today,* March 16, 2018.
9. Elizabeth C. Hirschman, "Cocaine as Innovation: A Social-Symbolic Account," in *NA—Advances in Consumer Research,* vol. 19, ed. John F. Sherry, Jr., and Brian Sternthal (Provo, Utah: Association for Consumer Research, 1992): 129–39; Andrew Golub and Bruce D. Johnson, "The Crack Epidemic: Empirical Findings Support an Hypothesized Diffusion of Innovation Process," *Socio-Economic Planning Sciences* 30, no. 3 (September 1996), 221–31.
10. Art Van Zee, "The Promotion and Marketing of OxyContin: Commercial Triumph, Public Health Tragedy," *American Journal of Public Health* 99, no. 2 (February 2009), 221–27.
11. Anushay Hossain, "Can an App Solve Racism? This Entrepreneur Says It Can," *Forbes,* September 5, 2016; Stephanie Marcus, "5 iPhone Apps to Help Fight Poverty," *Mashable,* September 16, 2010, https://mashable.com/2010/09/16/apps-fight-poverty/.
12. Robert Gordon, *The Rise and Fall of American Growth: The U.S. Standards of Living Since the Civil War* (Princeton: Princeton University Press, 2017). See also Nicholas Bloom, Charles I. Jones, John Van Reenen, and Michael Webb, "Are Ideas Getting Harder to Find?" *American Economic Review* (forthcoming), and Patrick Collison and Michael Nielsen, "Science Is Getting Less Bang for Its Buck," *The Atlantic,* November 16, 2018.
13. Guglielmo Mattioli, "What Caused the Genoa Bridge Collapse—and the End of an Italian National Myth?" *Guardian,* February 26, 2019.
14. Drake Baer, "Mark Zuckerberg Explains Why Facebook Doesn't 'Move Fast and Break Things' Anymore," *Business Insider,* May 2, 2014, https://www.businessinsider.com/mark-zuckerberg-on-facebooks-new-motto-2014-5.
15. One reason the idea resonated was because scholars in a variety of disciplines had published widely on the subjects of maintenance, infrastructure, and repair. Their work was and continues to be a source of vibrancy and inspiration. For a sampling, start with Ruth Schwartz Cowan, *More Work for Mother: The Ironies of Household Technology from the Open Hearth to the Microwave* (New York: Basic Books, 1983); Chris-

topher R. Henke, "The Mechanics of Workplace Order: Toward a Sociology of Repair," *Berkeley Journal of Sociology* 44 (1999–2000): 55–81; Pierre Claude Reynard, "Unreliable Mills: Maintenance Practices in Early Modern Papermaking," *Technology and Culture* 40, no. 2 (1999), 237–62; Stephen Graham and Nigel Thrift, "Out of Order: Understanding Repair and Maintenance," *Theory, Culture & Society* 24, no. 3 (2007), 1–25; Kevin L. Borg, *Auto Mechanics: Technology and Expertise in Twentieth-Century America* (Baltimore: Johns Hopkins University Press, 2007); David Edgerton, *The Shock of the Old: Technology and Global History Since 1900* (London: Profile Books, 2007); Steven J. Jackson, "Rethinking Repair," in *Media Technologies: Essays on Communication, Materiality, and Society,* ed. Tarleton Gillespie, Pablo Boczkowski, and Kirsten Foot (Cambridge, Mass.: MIT Press, 2014), 221–40; and Jérôme Denis and David Pontille, "Material Ordering and the Care of Things," *Science, Technology, & Human Values* 40, no. 3 (2015), 338–67; as well as the papers and presentations collected from the conferences and gatherings hosted by The Maintainers, available under the "Events" tab at themaintainers.org.

CHAPTER 2: TURNING ANXIETY INTO A PRODUCT

1. Christine MacLeod, *Heroes of Invention: Technology, Liberalism and British Identity, 1750–1914* (New York: Cambridge University Press, 2007); Joel Mokyr, *A Culture of Growth: The Origins of the Modern Economy* (Princeton, N.J.: Princeton University Press, 2016).
2. MacLeod, 1.
3. Angela Lakwete, *Inventing the Cotton Gin: Machine and Myth in Antebellum America* (Baltimore: Johns Hopkins University Press, 2005).
4. W. Patrick McCray, "It's Not All Lightbulbs," *Aeon,* October 12, 2016, https://aeon.co/essays/most-of-the-time-innovators-don-t-move-fast-and-break-things.
5. William S. Pretzer, "Introduction: The Meanings of the Two Menlo Parks," in *Working at Inventing: Thomas A. Edison and the Menlo Park Experience* (Baltimore: Johns Hopkins University Press, 2002), 12–31.
6. David C. Mowery and Nathan Rosenberg, *Paths of Innovation: Technological Change in 20th-Century America* (New York: Cambridge University Press, 1999), 4–5.
7. Citation figures come from Google Scholar. Solow, Robert M. "Technical Change and the Aggregate Production Function." *The Review of Economics and Statistics* (1957), 312–20.
8. U.S. Department of Commerce, Panel on Invention and Innovation,

Technological Innovation: Its Environment and Management (Washington, D.C.: U.S. Government Printing Office, 1967), 3, 81.
9. Daniel V. De Simone, *Education for Innovation* (Elsevier Science & Technology, 1968), 1.
10. De Simone, 2.
11. National Science Foundation, National Planning Association, *Proceedings of a Conference on Technology Transfer and Innovation* (Washington, D.C.: Government Printing Office, 1967).
12. Jill Lepore, "The Disruption Machine: What the Gospel of Innovation Gets Wrong," *New Yorker,* June 16, 2014.
13. Edward N. Wolff's *Top Heavy: The Increasing Inequality of Wealth in America and What Can Be Done about It* (New York: New Press, 1996) was an early work that noticed and examined rising inequality.
14. Chris Kirk and Will Oremus, "A World Map of All the 'Next Silicon Valleys,'" *Slate,* December 19, 2013, http://www.slate.com/articles/technology/the_next_silicon_valley/2013/12/all_the_next_silicon_valleys_a_world_map_of_aspiring_tech_hubs.html.
15. Lepore, "The Disruption Machine." See also Andrew A. King and Baljir Baatartogtokh, "How Useful Is the Theory of Disruptive Innovation?" *MIT Sloan Management Review* 57, no. 1 (Fall 2015), 77–90; Evan Goldstein, "The Undoing of Disruption," *Chronicle of Higher Education,* September 15, 2015; "The Myth of 'Disruptive Innovation,'" Robert H. Smith School of Business, September 15, 2015, https://www.rhsmith.umd.edu/news/myth-disruptive-innovation.
16. Richard Florida, *The Rise of the Creative Class, Revisited,* 10th anniversary edition (New York: Basic Books, 2012), 38.
17. The journalist Frank Bures has put together an excellent synthesis of critiques of Florida's "Creative Class" thesis: "Richard Florida Can't Let Go of His Creative Class Theory. His Reputation Depends on It," *BELT Magazine,* December 13, 2017, https://beltmag.com/richard-florida-cant-let-go/. See also Frank Bures, "The Fall of the Creative Class," *BELT Magazine,* June 15, 2012, https://beltmag.com/fall-of-the-creative-class.
18. Florida, 47.
19. Sam Wetherell, "Richard Florida Is Sorry," *Jacobin,* August 19, 2017, https://jacobinmag.com/2017/08/new-urban-crisis-review-richard-florida.
20. Tom Kelley and David Kelley, "Why Designers Need Empathy," *Slate,* November 8, 2013, https://slate.com/human-interest/2013/11/empathize-with-your-end-user-creative-confidence-by-tom-and-david-kelley.html.
21. Lilly Irani, "'Design Thinking': Defending Silicon Valley at the Apex of Global Labor Hierarchies," *Catalyst* 4, no. 1 (2018).

22. Peter N. Miller, "Is 'Design Thinking' the New Liberal Arts?" *The Chronicle of Higher Education,* March 26, 2015.
23. Natasha Jen, "Design Thinking Is Bullshit," 99U Conference 2017, https://99u.adobe.com/videos/55967/natasha-jen-design-thinking-is-bullshit.
24. Miller, "Is 'Design Thinking' the New Liberal Arts?"

CHAPTER 3: TECHNOLOGY AFTER INNOVATION

1. Brian X. Chen, "The Biggest Tech Failures and Successes of 2017," *New York Times,* December 13, 2017.
2. We are drawing on foundational accounts from David F. Noble, *America by Design: Science, Technology, and the Rise of Corporate Capitalism* (New York: Oxford University Press, 1979); Edwin T. Layton, *The Revolt of the Engineers: Social Responsibility and the American Engineering Profession* (Baltimore: Johns Hopkins University Press, 1986); Leo Marx, " 'Technology': The Emergence of a Hazardous Concept," *Social Research* 64, no. 4 (1997), 965–88; Ruth Oldenziel, *Making Technology Masculine: Men, Women and Modern Machines in America, 1870–1945* (Amsterdam: Amsterdam University Press, 1999); David Edgerton, *The Shock of the Old: Technology and Global History Since 1900* (London: Profile Books, 2007); Paul Nightingale, "What Is Technology? Six Definitions and Two Pathologies," SPRU—Science Policy Research Unit, University of Sussex Business School, 2014, https://ideas.repec.org/p/sru/ssewps/2014-19.html; Eric Schatzberg, *Technology: Critical History of a Concept* (Chicago: University of Chicago Press, 2018).
3. Ursula K. Le Guin, "A Rant about 'Technology,' " 2004, http://www.ursulakleguinarchive.com/Note-Technology.html.
4. Daniel Abramson, *Obsolescence: An Architectural History* (Chicago: University of Chicago Press, 2016).
5. Juliette Spertus and Valeria Mogilevich, "Super Strategies," *Urban Omnibus,* March 29, 2017, https://urbanomnibus.net/2017/03/super-strategies/; Jillian Steinhauer, "How Mierle Laderman Ukeles Turned Maintenance Work into Art," *Hyperallergic,* February 10, 2017, https://hyperallergic.com/355255/how-mierle-laderman-ukeles-turned-maintenance-work-into-art/.
6. Aryn Martin, Natasha Myers, and Ana Viseu, "The Politics of Care in Technoscience," *Social Studies of Science* 45, no. 5 (2015), 625–41; Michelle Murphy, "Unsettling Care: Troubling Transnational Itineraries of Care in Feminist Health Practices," *Social Studies of Science* 45, no. 5 (2015), 717–37; Sarah Leonard and Nancy Fraser, "Capitalism's Crisis of

Care," *Dissent* (Fall 2016), https://www.dissentmagazine.org/article/nancy-fraser-interview-capitalism-crisis-of-care.

7. U.S. Census 2010, *Population and Housing Unit Counts,* September 2012, table 10, https://www.census.gov/prod/cen2010/cph-2-1.pdf.
8. *Oxford English Dictionary,* s.v. "mechanic."
9. *Proceedings of the Roadmasters and Maintenance of Way Association* 24, 97–100.
10. Scott Reynolds Nelson, *Steel Drivin' Man: John Henry; The Untold Story of an American Legend* (New York: Oxford University Press, 2006), 109.
11. Kevin L. Borg, *Auto Mechanics: Technology and Expertise in Twentieth-Century America* (Baltimore: Johns Hopkins University Press, 2007).
12. Lawrence R. Dicksee, *Comparative Depreciation Tables* (London: Gee and Co., 1895); Ewing Matheson, *The Depreciation of Factories, Mines, and Industrial Undertakings and Their Valuation* (London: E. & F. N. Spon, 1893).
13. Federal Highway Administration, *Deferred Maintenance: Roadside Vegetation and Drainage Facilities*, report no. FHWA-RD-77-502 (August 1977), 1.
14. Federal Highway Administration, 40–1.
15. Robert Bond Randall, *Vibration-Based Condition Monitoring: Industrial, Aerospace and Automotive Applications* (Hoboken, N.J.: Wiley, 2011), xi.
16. "List of Vendors and Computerized Maintenance Management," in Terry Wireman, *Computerized Maintenance Management Systems* (New York: Industrial Press, 1986).

CHAPTER 4: SLOW DISASTER

1. Associated Press, "Feds: Poor Maintenance Led to Fatal DC Subway Fire," May 3, 2016.
2. NTSB, "Washington Metropolitan Area Transit Authority L'Enfant Plaza Station Electrical Arcing and Smoke Accident, Washington, D.C., January 12, 2015," NTSB/RAR-16/01, May 3, 2016, https://www.ntsb.gov/investigations/AccidentReports/Reports/RAR1601.pdf.
3. NTSB, "Ineffective Inspection, Maintenance Practices, Oversight Led to Washington Metrorail Fatal Accident," news release, May 3, 2016, https://www.ntsb.gov/news/press-releases/Pages/PR20160503.aspx.
4. Faiz Siddiqui, "Can Metro Trains Return to Automation? It's a $1 Million Question," *Washington Post,* June 9, 2018.
5. Justin George, "Returning to an Autopilot System Is Not in Metro's Plans for at Least Five Years, Safety Commission Says," *Washington Post,* December 10, 2019.

6. Robert McCartney and Paul Duggan, "Metro Sank into Crisis Despite Decades of Warnings," *Washington Post,* April 24, 2016.
7. Brian M. Rosenthal, Emma G. Fitzsimmons, and Michael LaForgia, "How Politics and Bad Decisions Starved New York's Subways," *New York Times,* November 18, 2017.
8. Jason Lange and Katanga Johnson, "Crumbling Bridges? Fret Not America, It's Not That Bad," Reuters, January 31, 2018, https://www.reuters.com/article/us-usa-trump-bridges/crumbling-bridges-fret-not-america-its-not-that-bad-idUSKBN1FK0J0.
9. Pat Choate and Susan Walter, *America in Ruins: The Decaying Infrastructure* (Durham, N.C.: Duke University Press, 1981), 1–3.
10. As quoted in Henry Petroski, *The Road Taken: The History and Future of America's Infrastructure* (New York: Bloomsbury, 2016), 15. This and the following two paragraphs draw heavily on Petroski's history of infrastructure reports.
11. National Council on Public Works Improvement, *Fragile Foundations: A Report on America's Public Works* (Washington, D.C., 1988), 7–8.
12. National Council on Public Works Improvement, 120.
13. National Council on Public Works Improvement, 6.
14. ASCE, "Report Card History," https://www.infrastructurereportcard.org/making-the-grade/report-card-history/.
15. ASCE, "Report Card History."
16. Charles Marohn, "My Journey from Free Market Ideologue to Strong Towns Advocate," Strong Towns blog, July 1, 2019, https://www.strongtowns.org/journal/2019/7/1/my-journey-from-free-market-ideologue-to-strong-towns-advocate.
17. Charles Marohn, "My Journey from Free Market Ideologue to Strong Towns Advocate."
18. Congressional Budget Office, *Public Spending on Transportation and Water Infrastructure, 1956 to 2014,* March 2015, 11, https://www.cbo.gov/sites/default/files/114th-congress-2015-2016/reports/49910-infrastructure.pdf.
19. Congressional Budget Office, 13.
20. Charles Marohn, "A Letter to POTUS on Infrastructure," Strong Towns blog, December 11, 2017, https://www.strongtowns.org/journal/2017/12/11/a-letter-to-potus-on-infrastructure.
21. Charles Marohn, "The Real Reason Your City Has No Money," Strong Towns blog, January 10, 2017, https://www.strongtowns.org/journal/2017/1/9/the-real-reason-your-city-has-no-money.
22. Charles Marohn, "Part 2: Mechanisms of Growth," Strong Towns blog, January 22, 2015, https://www.strongtowns.org/journal/2015/1/14/mechanisms-of-growth.

23. Charles Marohn, "Can You Be an Engineer and Speak Out for Reform?" Strong Towns blog, February 4, 2015, https://www.strongtowns.org/journal/2015/2/3/can-you-be-an-engineer-and-speak-out-for-reform.
24. Population estimates come from making inferences from M. B. Pell and Joshua Schneyer, "Thousands of U.S. Areas Afflicted with Lead Poisoning beyond Flint's," *Scientific American,* December 19, 2016.
25. Connor Sheets, "UN Poverty Official Touring Alabama's Black Belt: 'I Haven't Seen This' in the First World," AL.com, December 8, 2017, https://www.al.com/news/2017/12/un_poverty_official_touring_al.html.
26. Carlos Ballesteros, "Alabama Has the Worst Poverty in the Developed World, U.N. Official Says," *Newsweek,* December 10, 2017.
27. Ed Pilkington, "Hookworm, a Disease of Extreme Poverty, Is Thriving in the US South. Why?," *Guardian,* September 5, 2017.
28. Pilkington, "Hookworm."

CHAPTER 5: GROWTH AT ALL COSTS

1. George Bradt, "GE CEO Jeff Immelt's Long-Term View 10 Years In," *Forbes,* September 7, 2011; Jeffrey R. Immelt, "The Importance of Growth," GE Reports, June 17, 2015, https://www.ge.com/reports/post/121765814053/immelt-importance-of-growth/.
2. Steve Lohr, "G.E., the 124-Year-Old Software Start-Up," *New York Times,* August 27, 2016; Thomas Kellner, "GE Chairman and CEO Jeff Immelt's Annual Letter to GE Shareholders: 2014," GE Reports, March 16, 2015, https://www.ge.com/reports/post/113784948030/ge-chairman-and-ceo-jeff-immelts-annual-letter-to/.
3. Eli Cook, *The Pricing of Progress: Economic Indicators and the Capitalization of American Life* (Cambridge, Mass.: Harvard University Press, 2017), 16. On productivity and growth, see also Robert J. Gordon, *The Rise and Fall of American Growth: The U.S. Standard of Living since the Civil War* (Princeton, N.J.: Princeton University Press, 2016), and Robert M. Collins, *More: The Politics of Economic Growth in Postwar America* (New York: Oxford University Press, 2002).
4. Yessenia Funes, "California Power Company Tied to Last Year's Deadly Camp Fire Is Filing for Bankruptcy," Gizmodo, January 14, 2019, https://earther.gizmodo.com/california-power-company-tied-to-last-year-s-deadly-cam-1831733903; Raquel Maria Dillon, "Judge: PG&E Paid Out Stock Dividends Instead of Trimming Trees," KQED, April 2, 2019, https://www.kqed.org/news/11737336/judge-pge-paid-out-stock-dividends-instead-of-trimming-trees; Katherine Blunt and Russell

Gold, "PG&E Knew for Years Its Lines Could Spark Wildfires, and Didn't Fix Them," *Wall Street Journal,* July 10, 2019; J. D. Morris, "PG&E Is Less Than One-Third Done with Its 2019 Tree-Trimming Work," *San Francisco Chronicle,* October 1, 2019; Emma Newburger, " 'There Are Lives at Stake': PG&E Criticized over Blackouts to Prevent California Wildfires," CNBC, October 23, 2019, https://www.cnbc.com/2019/10/23/pge-rebuked-over-imposing-blackouts-in-california-to-reduce-fire-risk.html.

5. Cody Ogden, "Google Graveyard—Killed by Google," https://killedbygoogle.com/.
6. Natalie Kitroeff and David Gelles, "Claims of Shoddy Production Draw Scrutiny to a Second Boeing Jet," *New York Times,* April 20, 2019.
7. "Letter from Tim Cook to Apple Investors," Apple.com, January 2, 2019, https://www.apple.com/newsroom/2019/01/letter-from-tim-cook-to-apple-investors/.
8. Michael Sokolove, "How to Lose $850 Million—and Not Really Care," *New York Times,* June 9, 2002.
9. Peter Cohan, "Why Stack Ranking Worked Better at GE Than Microsoft," *Forbes,* July 13, 2012.
10. Drake Bennett, "How GE Went from American Icon to Astonishing Mess," *Bloomberg Businessweek,* February 1, 2018.
11. James B. Stewart, "Did the Jack Welch Model Sow Seeds of G.E.'s Decline?," *New York Times,* June 15, 2017; Jeff Spross, "The Fall of GE," *The Week,* March 19, 2018.
12. ASCE, "2017 Infrastructure Report Card: Schools," https://www.infrastructurereportcard.org/cat-item/schools/.
13. "Moody's—US Higher Education Outlook Remains Negative on Low Tuition Revenue Growth," Moody's, December 5, 2018, https://www.moodys.com/research/Moodys-US-higher-education-outlook-remains-negative-on-low-tuition—PBM_1152326.
14. See Christopher Newfield, *The Great Mistake: How We Wrecked Public Universities and How We Can Fix Them* (Baltimore: Johns Hopkins University Press, 2016); Elizbeth Popp Berman, *Creating the Market University: How Academic Science Became an Economic Engine* (Princeton, N.J.: Princeton University Press, 2011); Paul Nightingale and Alex Coad, "The Myth of the Science Park Economy," *Demos Quarterly,* issue 2 (Spring 2014).
15. Matthew Lynch, "Chronicling the Biggest EdTech Failures of the Last Decade," Tech Advocate, July 10, 2019, https://www.thetechedvocate.org/chronicling-the-biggest-edtech-failures-of-the-last-decade/.
16. Jill Barshay, "Research Shows Lower Test Scores for Fourth Graders

Who Use Tablets in Schools," Hechinger Report, June 10, 2019, https://hechingerreport.org/research-shows-lower-test-scores-for-fourth-graders-who-use-tablets-in-schools/.

17. Christo Sims, "How Idealistic High-Tech Schools Often Fail to Help Poor Kids Get Ahead," Zócalo Public Square, June 13, 2019, https://www.zocalopublicsquare.org/2019/06/13/how-idealistic-high-tech-schools-often-fail-to-help-poor-kids-get-ahead/ideas/essay/.
18. Christo Sims, *Disruptive Fixation: School Reform and the Pitfalls of Techno-Idealism* (Princeton, N.J.: Princeton University Press 2017), 11; Audrey Watters, "The 100 Worst Ed-Tech Debacles of the Decade," Hack Education, December 31, 2019, http://hackeducation.com/2019/12/31/what-a-shitshow.
19. Morgan G. Ames, *The Charisma Machine: The Life, Death, and Legacy of One Laptop per Child* (Cambridge, Mass.: MIT Press, 2019); Marc Tracy and Tiffany Hsu, "Director of M.I.T.'s Media Lab Resigns After Taking Money from Jeffrey Epstein," *New York Times,* September 7, 2019.
20. Roderic N. Crooks, "The Coded Schoolhouse: One-to-One Tablet Computer Programs and Urban Education" (PhD diss., UCLA, 2016).
21. Max Roser, Hannah Ritchie, and Bernadeta Dadonaite, "Child & Infant Mortality," Our World in Data, 2019, https://ourworldindata.org/child-mortality.
22. Lisa Rapaport, "U.S. Health Spending Twice Other Countries' with Worse Results," Reuters, March 13, 2018, https://www.reuters.com/article/us-health-spending/u-s-health-spending-twice-other-countries-with-worse-results-idUSKCN1GP2YN.
23. FDA, Office of Orphan Products Development, Orphan Drug Designation and Approval Database, https://www.accessdata.fda.gov/scripts/opdlisting/oopd/.

CHAPTER 6: THE MAINTAINER CASTE

1. Deborah M. Gordon, "Dynamics of Task Switching in Harvester Ants," *Animal Behaviour* 38, no. 2 (1989): 194–204.
2. Anuradha Nagaraj, "Activist Helping Lower Castes in India Forced to Clean Toilet Feces by Hand," *HuffPost,* July 28, 2016, https://www.huffpost.com/entry/activist-helping-lower-castes-in-india-forced-to-clean-toilet-feces-by-hand_n_579a28b7e4b02d5d5ed4ab7d.
3. Associated Press, "The 'Untouchables' of Yemen Caught in Crossfire of War," Fox News, May 17, 2016, https://www.foxnews.com/world/the-untouchables-of-yemen-caught-in-crossfire-of-war.
4. Melanie Mills, Shirley K. Drew, and Bob M. Gassaway, introduction to

Dirty Work: The Social Construction of Taint (Waco, Tex.: Baylor University Press, 2007), 1.

5. Kevin L. Borg, *Auto Mechanics: Technology and Expertise in Twentieth-Century America* (Baltimore: Johns Hopkins University Press, 2007).
6. John Levi Martin, "What Do Animals Do All Day?: The Division of Labor, Class Bodies, and Totemic Thinking in the Popular Imagination," *Poetics* 27 (2000), 195–231.
7. Carl Hendrick, "Why Schools Should Not Teach General Critical-Thinking Skills," *Aeon,* December 5, 2016, https://aeon.co/ideas/why-schools-should-not-teach-general-critical-thinking-skills.
8. David Edgerton, *The Shock of the Old: Technology and Global History Since 1900* (New York: Oxford University Press, 2011).
9. Barry Boehm, an early authority on software engineering, published a study in 1976 that included data from General Motors, the telecommunications firm GTE, and two U.S. Air Force units. Together, they reported that maintenance was responsible for between 60 and 80 percent of the overall costs of software; see B. W. Boehm, "Software Engineering," *IEEE Transactions on Computers* 25, no. 12 (1976), 1226–41.
10. Trent Hamm, "Why You Should Consider Trade School Instead of College," The Simple Dollar, January 24, 2019, https://www.thesimpledollar.com/investing/college/why-you-should-consider-trade-school-instead-of-college/.
11. Email from Melinda Hodkiewicz, June 5, 2017.
12. Matthew Yglesias, "The 'Skills Gap' Was a Lie," Vox, January 7, 2019, https://www.vox.com/2019/1/7/18166951/skills-gap-modestino-shoag-ballance.
13. Borg, *Auto Mechanics,* 5.
14. Kari Paul, "Division of Labor Is a Big Problem at Work: Women Are Asked to Do 'Office Housework' by Their Male Co-workers," MarketWatch, May 12, 2019, https://www.marketwatch.com/story/already-paid-less-than-men-women-are-still-asked-to-do-the-office-housework-2018-10-08.
15. Charles Taylor, "The Politics of Recognition," in *Multiculturalism: Examining the Politics of Recognition,* ed. Amy Gutmann (Princeton, N.J.: Princeton University Press, 1994), 25–73.
16. Verónica Caridad Rabelo and Ramaswami Mahalingam, "'They Really Don't Want to See Us': How Cleaners Experience Invisible 'Dirty' Work," *Journal of Vocational Behavior* 113 (2019), 103–14.
17. "2019 Health & Human Services Poverty Guidelines," Paying for Senior Care, May 2019, https://www.payingforseniorcare.com/longtermcare/federal-poverty-level.html.

18. "What Is the Current Poverty Rate in the United States?" Center for Poverty Research, University of California, Davis, October 15, 2018, https://poverty.ucdavis.edu/faq/what-current-poverty-rate-united-states.

CHAPTER 7: A CRISIS OF CARE

1. YouGov, September 10, 2013, http://cdn.yougov.com/cumulus_uploads/document/ypg8eyjbsv/tabs_skincare_0910112013.pdf.
2. Kaitlyn McLintock, "The Average Cost of Beauty Maintenance Could Put You through Harvard," Byrdie, June 26, 2017, https://www.byrdie.com/average-cost-of-beauty-maintenance.
3. Cass R. Sunstein, "It Captures Your Mind," *New York Review of Books,* September 26, 2013.
4. "Weight Management," Boston Medical Center, https://www.bmc.org/nutrition-and-weight-management/weight-management; Michael Hobbes, "Everything You Know about Obesity Is Wrong," *Highline,* September 19, 2018, https://highline.huffingtonpost.com/articles/en/everything-you-know-about-obesity-is-wrong/.
5. Evelyn Nakano Glenn, *Forced to Care: Coercion and Caregiving in America* (Cambridge, Mass.: Harvard University Press, 2012), 2; A. W. Geiger, Gretchen Livingston, and Kristen Bialik, "6 Facts about U.S. Moms," Pew Research Center, May 8, 2019, https://www.pewresearch.org/fact-tank/2019/05/08/facts-about-u-s-mothers/.
6. Glenn, 3.
7. Glenn, 4.
8. Nancy Fraser, "Contradictions of Capital and Care," *New Left Review* 100 (2016).
9. Claire Cain Miller, "How Same-Sex Couples Divide Chores, and What It Reveals about Modern Parenting," *New York Times,* May 16, 2018.
10. "Average Size of US Homes, Decade by Decade," Newser, May 29, 2016, https://www.newser.com/story/225645/average-size-of-us-homes-decade-by-decade.html.
11. Jessica Guerin, "Americans Are Way More in Debt Now Than They Were after the Financial Crisis," HousingWire, February 12, 2019, https://www.housingwire.com/articles/48162-americans-are-way-more-in-debt-now-than-they-were-after-the-financial-crisis.
12. Veronica Mosqueda and Rob Wohl, "A Columbia Heights Rent Strike Highlights Abuses Low-Income Tenants Face in DC," Greater Greater Washington, April 3, 2019, https://ggwash.org/view/71558/a-columbia-heights-rent-strike-highlights-abuses-tenants-face-in-dc.
13. Rosalind Williams in *The Durability Factor: A Guide to Finding Long-*

Lasting Cars, Housing, Clothing, Appliances, Tools, and Toys, ed. Roger B. Yepsen, Jr. (Emmaus, PA: Rodale Press, 1982), 12.

14. K. E. McFadden, "Garagecraft: Tinkering in the American Garage" (PhD diss., University of South Carolina, 2018), 31.
15. Patrick Sisson, "Self-Storage: How Warehouses for Personal Junk Became a $38 Billion Industry," Curbed, March 27, 2018, https://www.curbed.com/2018/3/27/17168088/cheap-storage-warehouse-self-storage-real-estate.
16. Arielle Bernstein, "Marie Kondo and the Privilege of Clutter," *Atlantic,* March 25, 2016.
17. Antonio Villas-Boas, "Apple Quoted Me $1,500 to Repair a MacBook Pro, So I Paid Less Than $500 at an 'Unauthorized' Apple Repair Shop Instead," *Business Insider,* December 16, 2018, https://www.businessinsider.com/apple-macbook-pro-repair-quote-unauthorized-2018-12.
18. Kyle Wiens, "The New MacBook Pro: Unfixable, Unhackable, Untenable," *Wired,* June 14, 2012; Caroline Haskins, "AirPods Are a Tragedy," *Motherboard,* May 6, 2019, https://www.vice.com/en_us/article/neaz3d/airpods-are-a-tragedy.
19. Jason Farman, "Repair and Software: Updates, Obsolescence, and Mobile Culture's Operating Systems," *Continent* 6.1 (2017), http://www.continentcontinent.cc/index.php/continent/article/view/275.

CHAPTER 8: THE MAINTENANCE MINDSET

1. Wei Lin Koo and Tracy Van Hoy, "Determining the Economic Value of Preventive Maintenance," Jones Lang LaSalle, https://gridium.com/wp-content/uploads/economic-value-of-preventative-maintenance.pdf.
2. Augury, "Case Study: Large Home Appliance, Refrigerator Manufacturing Facility," http://info.augury.com/Appliance-Manufacturing-Case-Study-WEB-pdf.html, and Augury, "Case Study: Medical Device Manufacturing Facility," http://info.augury.com/Hologic-Case-Study-Augury-pdf.html.
3. Andrea Goulet, email to Andrew Russell, November 16, 2017.
4. Netflix Technology Blog, "The Netflix Simian Army," Medium, July 19, 2011, https://medium.com/netflix-techblog/the-netflix-simian-army-16e57fbab116. See also "The Origin of Chaos Monkey: Why Netflix Needed to Create Failure," Gremlin, October 16, 2018, https://www.gremlin.com/chaos-monkey/the-origin-of-chaos-monkey/.
5. Congressional Budget Office, "Trends in Spending by the Department of Defense for Operation and Maintenance," January 5, 2017, https://www.cbo.gov/publication/52156.

6. Nicolas Niarchos, "How the U.S. Is Making the War in Yemen Worse," *New Yorker,* January 15, 2018.
7. Congressional Budget Office, "The Depot-Level Maintenance of DoD's Combat Aircraft: Insights for the F-35," February 16, 2018, https://www.cbo.gov/publication/53543; Congressional Budget Office, "Trends in Spending by the Department of Defense for Operation and Maintenance."
8. ISO 55000:2014(en) "Asset Management—Overview, Principles and Terminology," https://www.iso.org/obp/ui/#iso:std:iso:55000:ed-1:v2:en.
9. "Corporate Social Responsibility," Fiix, https://www.fiixsoftware.com/csr/; Craig Daniels, "How One CEO Hardwired His Company for Good," Communitech News, May 24, 2018, https://news.communitech.ca/how-one-ceo-hardwired-his-company-for-good/.
10. Julie E. Wollman, "A Burst Pipe Brings a Flood of Insights for a University President," *Chronicle of Higher Education,* April 23, 2019.
11. "Computerized Maintenance Management System (CMMS) Software Market 2019 Global Industry—Key Players, Size, Trends, Opportunities, Growth Analysis and Forcecast to 2025," press release, MarketWatch, February 7, 2019, https://www.marketwatch.com/press-release/computerized-maintenance-management-system-cmms-software-market-2019-global-industry---key-players-size-trends-opportunities-growth-analysis-and-forecast-to-2025-2019-02-07.
12. Grand View Research, "Industrial IoT Market Size Worth $949.42 Billion by 2025," June 2019, https://www.grandviewresearch.com/press-release/global-industrial-internet-of-things-iiot-market.
13. Steve Lohr, "G.E., the 124-Year-Old Software Start-Up," *New York Times,* August 27, 2016.
14. "Bringing Maintenance into the Fourth Industrial Revolution," *Manufacturers' Monthly,* December 17, 2018.
15. "Business Roundtable Supports Move Away from Short-Term Guidance," https://www.businessroundtable.org/archive/media/news-releases/business-roundtable-supports-move-away-short-term-guidance.

CHAPTER 9: FIX IT FIRST

1. Kim Dacey, "Report Says Water Affordability Is Race Issue in Baltimore," WBAL-TV, June 12, 2019, https://www.wbaltv.com/article/water-affordability-a-race-issue-in-baltimore-report/27921606#.
2. Daniel Bush, "What Would It Take to Fix America's Crumbling Infrastructure?" *PBS NewsHour,* January 8, 2018, https://www.pbs.org

/newshour/economy/making-sense/what-would-it-take-to-fix-americas-crumbling-infrastructure.

3. Bipartisan Policy Center, "Bridging the Gap Together: A New Model to Modernize U.S. Infrastructure," May 2016, https://bipartisanpolicy.org/wp-content/uploads/2019/03/BPC-New-Infrastructure-Model.pdf.
4. Bipartisan Policy Center, 44.
5. Jill Eicher, "Some Love for the Infrastructure We Already Have," Governing, February 4, 2019, https://www.governing.com/gov-institute/voices/col-infrastructure-deferred-maintenance-balance-sheets-financial-reports.html.
6. Charles L. Marohn, "Misunderstanding Mobility" in *Thoughts on Building Strong Towns,* vol. 1 (CreateSpace Independent Publishing Platform, 2012), 48–75.
7. Marc Reisner, *Cadillac Desert: The American West and Its Disappearing Water* (New York: Penguin, 1993), 3.
8. Seki, "Managing Maintenance on the Tokaido Shinkansen," *Railway Gazette International,* August 1, 2003.
9. Christopher Ingraham, "The Sorry State of Amtrak's On-Time Performance, Mapped," *Washington Post,* July 10, 2014.
10. Amtrak, *Amtrak Five Year Equipment Asset Line Plan: Base (FY 2019) + Five Year Strategic Plan (FY 2020–2024),* https://www.amtrak.com/content/dam/projects/dotcom/english/public/documents/corporate/businessplanning/Amtrak-Equipment-Asset-Line-Plan-FY20-24.pdf.
11. Richard Medhurst, "Doctor Yellow Keeps the Shinkansen Network Healthy," Nippon.com, April 28, 2016, https://www.nippon.com/en/nipponblog/m00107/doctor-yellow-keeps-the-shinkansen-network-healthy.html.
12. Chris Iovenko, "Dutch Masters: The Netherlands Exports Flood-Control Expertise," *EARTH,* October 15, 2018, https://www.earthmagazine.org/article/dutch-masters-netherlands-exports-flood-control-expertise.
13. "Dutch Dialogues: New Orleans," Waggonner & Ball, https://wbae.com/projects/dutch_dialogues_new_orleans.
14. Bush, "What Would It Take to Fix America's Crumbling Infrastructure?"
15. "Crumbling Infrastructure Is a Worldwide Problem," *Economist,* August 18, 2018.
16. Bobby Allyn and Frank Langfitt, "Leaked Brexit Document Depicts Government Fears of Gridlock, Food Shortages, Unrest," NPR, August 18, 2019, https://www.npr.org/2019/08/18/752173091/leaked-brexit-document-depicts-government-fears-of-gridlock-food-shortages-unrest.

17. New York Post Editorial Board, "The $2B Lunacy of the LaGuardia AirTrain," *New York Post,* July 1, 2019.
18. TransitCenter, "Who's on Board 2016: What Today's Riders Teach Us about Transit That Works," November 21, 2016, https://transitcenter.org/publication/whos-on-board-2016/.
19. "Bus Turnaround: 2018: Fast Bus, Fair City," BusTurnaround.nyc, http://busturnaroundn.wpengine.com/wp-content/uploads/2018/07/BusTurnaroundAction-Plan.pdf.
20. TransitCenter, "Getting to the Route of It: The Role of Governance in Regional Transit," October 9, 2014, https://transitcenter.org/getting-to-the-route-of-it/.
21. TransitCenter, "Getting to the Route of It."
22. TransitCenter, "Getting to the Route of It."

CHAPTER 10: SUPPORTING THE WORK THAT MATTERS MOST

1. Robert N. Charette, "The STEM Crisis Is a Myth," *IEEE Spectrum,* August 30, 2013.
2. Gwen Burrow, "The Most Popular Jobs for Young Workers Are in the Arts, Skilled Trades, and Sciences," Emsi, May 4, 2017, https://www.economicmodeling.com/2017/05/04/best-jobs-workers-25/; Matt Krupnick, "After Decades of Pushing Bachelor's Degrees, U.S. Needs More Tradespeople," *PBS NewsHour,* August 29, 2017, https://www.pbs.org/newshour/education/decades-pushing-bachelors-degrees-u-s-needs-tradespeople.
3. Livia Gershon, "The Future Is Emotional," *Aeon,* June 22, 2017, https://aeon.co/essays/the-key-to-jobs-in-the-future-is-not-college-but-compassion; David J. Deming, "The Growing Importance of Social Skills in the Labor Market," NBER Working Paper No. 21473, August 2015, https://www.nber.org/papers/w21473.
4. Mike Rowe, "Learning from Dirty Jobs," TED, December 2008, https://www.ted.com/talks/mike_rowe_celebrates_dirty_jobs/.
5. The Kitchen Sisters, "The Working Tapes of Studs Terkel," http://www.kitchensisters.org/present/the-working-tapes-of-studs-terkel/. For criticisms of Rowe's funding sources and political positions see, for example, Nima Shirazi and Adam Johnson, "Episode 64: Mike Rowe's Koch-Backed Working Man Affectation," Citations Needed, January 30, 2019, https://medium.com/@CitationsPodcst/episode-64-mike-rowes-koch-backed-working-man-affectation-fa52e0e8d2e3.
6. Rick Berger, "All the Ways the US Military's Infrastructure Crisis Is Getting Worse," *Defense One,* March 27, 2019, https://www.defenseone

.com/ideas/2019/03/us-militarys-infrastructure-crisis-only-getting-worse/155858/; Stephen Losey, "Fewer Planes Are Ready to Fly: Air Force Mission-Capable Rates Decline amid Pilot Crisis," *Air Force Times,* March 5, 2018.

7. Millen Paschich, "Color-Coding Complex Maintenance," Gridium, July 17, 2017, https://gridium.com/color-coding-complex-maintenance/.
8. U.S. Air Force, "Maintainers: The Driving Force," November 10, 2016, https://www.af.mil/News/Article-Display/Article/1001950/maintainers-the-driving-force/.
9. Oriana Pawlyk, "The Air Force Has Fixed Its Active-Duty Maintainer Shortage, SecAF Says," Military.com, February 8, 2019, https://www.military.com/dodbuzz/2019/02/08/air-force-has-fixed-its-active-duty-maintainer-shortage-secaf-says.html. See also U.S. Department of Defense, "Aircraft Maintainers: Lifelines of the Air Force," January 11, 2019, https://www.defense.gov/explore/story/Article/1729504/aircraft-maintainers-lifelines-of-the-air-force/.
10. Government Accountability Office, "Military Personnel: Strategy Needed to Improve Retention of Experienced Air Force Aircraft Maintainers," February 2019, https://www.gao.gov/assets/700/696747.pdf; Jennifer H. Svan, "Air Force Needs to Do More to Keep Experienced Maintainers, Report Says," *Stars and Stripes,* February 7, 2019; Stephen Losey, "The Air Force Still Has a Serious Maintainer Staffing Problem, GAO Says—but No Strategy to Fix It," *Air Force Times,* February 8, 2019.
11. See also u/H0stusM0stus, "One Maintainers Opinion on Why Maintainers Are Not Staying In," Reddit, subreddit r/AirForce, https://www.reddit.com/r/AirForce/comments/3tez9v/one_maintainers_opinion_on_why_maintainers_are/.
12. See *The Moderators,* directed by Ciaran Cassidy and Adrian Chen (2017), https://vimeo.com/239108604; Tarleton Gillespie, *Custodians of the Internet: Platforms, Content Moderation, and the Hidden Decisions That Shape Social Media* (New Haven, Conn.: Yale University Press, 2018); Sarah T. Roberts, *Behind the Screen: Content Moderation in the Shadows of Social Media* (New Haven, Conn.: Yale University Press, 2019); Mary L. Gray and Siddharth Suri, *Ghost Work: How to Stop Silicon Valley from Building a New Global Underclass* (Boston: Houghton Mifflin Harcourt, 2019).
13. Daisuke Wakabayashi, "Google's Shadow Work Force: Temps Who Outnumber Full-Time Employees," *New York Times,* May 28, 2019; Louis Hyman, *Temp: The Real Story of What Happened to Your Salary, Benefits, and Job Security* (New York: Penguin, 2019).
14. American Association of Colleges of Nursing, "Fact Sheet: Nursing

Shortage," April 2019, https://www.aacnnursing.org/Portals/42/News/Factsheets/Nursing-Shortage-Factsheet.pdf.

15. "Nurses Change Lives," Johnson & Johnson, https://nursing.jnj.com/.
16. Emily Beater, "Social Care Robots Privatise Loneliness, and Erode the Pleasure of Being Truly Known," *New Statesman America,* August 7, 2019. The article features an interview with a human healthcare worker named Casey: "The lady whom Casey cares for likes it when she pours her a glass of wine, or when she looks in the freezer and notices that she's running low on her favourite ice-cream. A home health robot would be useful. The client would never run out of ice-cream, the robot would pre-order it and the freezer would always be fully stocked. But the client would miss out on Casey remembering her; on the pleasure of another human being knowing her."
17. Nolan Lawson, "What It Feels Like to Be an Open-Source Maintainer," March 5, 2017, https://nolanlawson.com/2017/03/05/what-it-feels-like-to-be-an-open-source-maintainer/. For a more general treatment of the topic, see Nadia Eghbal, *Roads and Bridges: The Unseen Labor behind Our Digital Infrastructure,* Ford Foundation, 2016, https://www.fordfoundation.org/media/2976/roads-and-bridges-the-unseen-labor-behind-our-digital-infrastructure.pdf.
18. Jess Frazelle, "Customer Story," GitHub, https://github.com/customer-stories/jessfraz.
19. Jan Lehnardt, "Sustainable Open Source: The Maintainers Perspective; or How I Learned to Stop Caring and Love Open Source," March 6, 2017, https://writing.jan.io/2017/03/06/sustainable-open-source-the-maintainers-perspective-or-how-i-learned-to-stop-caring-and-love-open-source.html.
20. Ariya Hidayat, "Customer Story," GitHub, https://github.com/customer-stories/ariya; Henry Zhu, "Customer Story," GitHub, https://github.com/customer-stories/hzoo.
21. Linus Torvalds and David Diamond, *Just for Fun: The Story of an Accidental Revolutionary* (New York: HarperCollins, 2001), 238.

CHAPTER 11: CARING FOR OUR HOMES, OUR STUFF, AND ONE ANOTHER

1. Justin Ward, "Don't Toss It, Fix It: Habitat for Humanity Brings Repair Cafe to NRV," WDBJ7, January 26, 2017, https://www.wdbj7.com/content/news/Dont-toss-it-fix-it-Habitat-for-Humanity-brings-Repair-Cafe-to-NRV-411924655.html.
2. Tonia Moxley, "Repair Cafe Hopes to Combat Today's Throw-Away Mindset," *Roanoke Times,* October 14, 2017.

3. The quotations in this and the following paragraph come from Tonia Moxley, "First Tool 'Library' Meant to Build Community Self-Reliance," *Roanoke Times,* October 8, 2018.
4. Haley Stewart, *The Grace of Enough: Pursuing Less and Living More in a Throwaway Culture* (Notre Dame, IN: Ave Maria Press, 2018), xviii.
5. Ruth Schwartz Cowan, *More Work for Mother: The Ironies of Household Technology from the Open Hearth to the Microwave* (New York: Basic Books, 1983), 216.
6. Cowan, *More Work for Mother,* 218.
7. Cowan, *More Work for Mother,* 219.
8. Cowan, *More Work for Mother,* 219.
9. Nicholas Gerbis, "How Much Does Auto Maintenance Cost over Time?," HowStuffWorks, https://auto.howstuffworks.com/under-the-hood/cost-of-car-ownership/auto-maintenance-cost.htm.
10. Zack Friedman, "78% of Workers Live Paycheck to Paycheck," *Forbes,* January 11, 2019.
11. Maurie Backman, "Nearly 3 in 5 Americans Are Making This Huge Financial Mistake," CNN Money, October 24, 2016, https://money.cnn.com/2016/10/24/pf/financial-mistake-budget/index.html.
12. iFixit, "Introducing Repair Tips from the Fixit Clinic with Peter Mui," YouTube, July 18, 2019, https://www.youtube.com/watch?v=1IJwpFBmTGk.
13. Peter Mui, "Celebrating Repair at Fixit Clinic," iFixit, July 11, 2016, https://www.ifixit.com/News/celebrating-repair-fixit-clinic.
14. Jason Koebler, "Apple Tells Lawmaker That Right to Repair iPhones Will Turn Nebraska into a 'Mecca' for Hackers," *Vice,* February 17, 2017, https://www.vice.com/en_us/article/pgxgpg/apple-tells-lawmaker-that-right-to-repair-iphones-will-turn-nebraska-into-a-mecca-for-hackers.
15. Claire Bushey, "Why Deere and Cat Don't Want Customers to Do It Themselves," *Crain's Chicago Business,* May 10, 2019; Antonio Villas-Boas, "Apple Quoted Me $1,500 to Repair a MacBook Pro, So I Paid Less Than $500 at an 'Unauthorized' Apple Repair Shop Instead," *Business Insider,* December 16, 2018, https://www.businessinsider.com/apple-macbook-pro-repair-quote-unauthorized-2018-12.
16. Richard Orange, "Waste Not Want Not: Sweden to Give Tax Breaks for Repairs," *Guardian,* September 19, 2016.

EPILOGUE: FROM CONVERSATION TO ACTION

1. Larry Neal, "The Black Arts Movement," *Drama Review* (Summer 1968), at nationalhumanitiescenter.org/pds/maai3/community/text8/blackartsmovement.pdf.

2. Audre Lorde, *A Burst of Light* (Ithaca, NY: Firebrand Books, 1988), 130.
3. "Why Regenerative Agriculture?" Regeneration International, https://regenerationinternational.org/why-regenerative-agriculture/.
4. Gracy Olmstead, "The Perils of Innovation," *American Conservative,* April 15, 2016.
5. Sean Captain, "How a Socialist Coder Became a Voice of Engineers Standing Up to Management," *Fast Company,* October 15, 2018.

Index

About the Authors

Lee Vinsel is a professor in the Department of Science, Technology, and Society at Virginia Tech. Andrew Russell is a professor of history and dean of the College of Arts and Sciences at SUNY Polytechnic Institute. Together, they are the founders of the Maintainers research network and conferences, and their writing on the topics of this book has appeared in or been covered by *The New York Times, The Atlantic, The Washington Post,* and *Wired*.

About the Type

This book was set in Granjon, a modern recutting of a typeface produced under the direction of George W. Jones (1860–1942), who based Granjon's design upon the letterforms of Claude Garamond (1480–1561). The name was given to the typeface as a tribute to the typographic designer Robert Granjon (1513–89).